ANXIETY

ANXIETY

❖ ❖ ❖

A Short History

Allan V. Horwitz

THE JOHNS HOPKINS UNIVERSITY PRESS
Baltimore

© 2013 The Johns Hopkins University Press
All rights reserved. Published 2013
Printed in the United States of America on acid-free paper

2 4 6 8 9 7 5 3 1

The Johns Hopkins University Press
2715 North Charles Street
Baltimore, Maryland 21218-4363
www.press.jhu.edu

Library of Congress Cataloging-in-Publication Data
Horwitz, Allan V.
Anxiety : a short history / Allan V. Horwitz.
p. ; cm. — (Johns Hopkins biographies of disease)
Includes bibliographical references and index.
ISBN-13: 978-1-4214-1080-7 (pbk. : alk. paper)
ISBN-10: 1-4214-1080-X (pbk. : alk. paper)
ISBN-13: 978-1-4214-1081-4 (electronic)
ISBN-10: 1-4214-1081-8 (electronic)
I. Title. II. Series: Johns Hopkins biographies of disease.
[DNLM: 1. Anxiety Disorders—history. 2. Anxiety—history. WM 11.1]
RC531
616.85'22—dc23
2013003541

A catalog record for this book is available from the British Library.

Special discounts are available for bulk purchases of this book. For more information,
please contact Special Sales at 410-516-6936 or specialsales@press.jhu.edu.

The Johns Hopkins University Press uses environmentally friendly book materials,
including recycled text paper that is composed of at least 30 percent post-consumer waste,
whenever possible.

To Jane

CONTENTS

Disease is a fundamental aspect of the human condition. Ancient bones tell us that pathological processes are older than humankind's written records, and sickness and death still confound us. We have not banished pain, disability, or the fear of death, even though we now die on average at older ages, of chronic more often than acute ills, and in hospital or hospice beds and not in our own homes. Disease is something men and women feel. It is experienced in our bodies—but also in our minds and emotions. It can bring pain and incapacity and hinder us at work and in meeting family responsibilities. Disease demands explanation; we think about it and we think with it. Why have I become ill? And why now? How is my body different in sickness from its quiet and unobtrusive functioning in health? Why in times of epidemic has a community been scourged?

Answers to such timeless questions necessarily mirror and incorporate time- and place-specific ideas, social assumptions, and technological options. In this sense, disease has always been a social and linguistic, a cultural as well as biological, entity. In the Hippocratic era more than two thousand years ago, physicians (we have always had them with us) were limited to the evidence of their senses in diagnosing a fever, an abnormal discharge, or seizure. Their notions of the material basis for such felt and visible symptoms necessarily reflected and incorporated contemporary philosophical and physiological notions, a speculative world of disordered humors, "breath," and pathogenic local environments. Today we can call upon a rather different variety of scientific understandings and an armory of diagnostic practices—tools that allow us to diagnose ailments unfelt by patients and imperceptible to the doctor's unaided senses. In the past century disease has also become increasingly a bureaucratic phenomenon, as sickness has been defined—and in that sense constituted—by formal disease classifications, treatment protocols, and laboratory thresholds.

Sickness is also linked to climatic and geographic factors. How and where we live and how we distribute our resources contribute to the incidence of disease. For example, ailments such as typhus fever, plague, malaria, dengue, and yellow fever reflect specific environments that we have shared with our insect contemporaries. But humankind's physical circumstances are determined in part by culture, and especially by agricultural practice in the millennia before the growth of cities and industry. What we eat and the work we do or do not do—our physical as well as cultural environment—all help determine our health and longevity. Environment, demography, economic circumstances, and applied medical knowledge all interact to create particular distributions of disease at particular places and specific moments in time. The twenty-first-century ecology of sickness in the developed world is marked, for example, by the dominance of chronic and degenerative illnesses—ailments of the cardiovascular system, of the kidneys, and cancer.

Disease is historically as well as ecologically specific. Or perhaps I should say that every disease has a unique past. Once discerned and named, every disease claims its own history. At one level, biology creates that identity. Symptoms and epidemiology, but also generation-specific cultural values and scientific understandings, shape our responses to illness. Some writers may have romanticized tuberculosis—think of Greta Garbo as Camille— but, as the distinguished medical historian Owsei Temkin noted dryly, no one ever thought to romanticize dysentery. Tuberculosis was pervasive in nineteenth-century Europe and North America and killed far more women and men than cholera did, but it never mobilized the same widespread and policy-shifting concern. It was a familiar aspect of life, to be endured if not precisely accepted. Unlike tuberculosis, cholera killed quickly and dramatically; it was never accepted as a condition of life in Europe and North America. Its episodic visits were anticipated with fear. Sporadic cases of influenza are normally invisible, indistinguishable among a variety of respiratory infections; waves of epidemic flu are all too visible. Syphilis and other sexually transmitted diseases, to cite another example, have had a peculiar and morally inflected history. Some diseases, such as smallpox and malaria, have a long history; others, like AIDS, a rather short one. Some diseases, like diabetes and cardiovascular disorders, have flourished in modern circumstances; others reflect the realities of an earlier and economically less developed world.

These arguments constitute the logic motivating and underlying the Johns Hopkins Biographies of Disease. *Biography* implies an identity, a chronology, and a narrative—a movement in and through time. Once inscribed by name in our collective understanding of medicine and the body, each disease entity becomes a part of that collective understanding and thus inevitably shapes the way individual men and women think about their own felt symptoms and prospects for future health. Each historically visible entity—each disease—has a distinct history, even if that history is not always defined in terms familiar to twenty-first-century physicians. The very notion of specific disease entities—fixed and based on a defining mechanism—is in itself a historical artifact. "Dropsy" and "Bright's disease" are no longer terms of everyday clinical practice, but they are an unavoidable part of the history of chronic kidney disease. Nor do we speak of "essential," "continued," "bilious," or "remittent" fevers; they are not meaningful designations in today's disease classifications. Fever is now a symptom, the body's response to a triggering circumstance. It is no longer a "disease," as it was through millennia of human history. "Flux," or diarrhea, is similarly no longer a disease but a symptom associated with a variety of specific and nonspecific causes. We have come to assume and expect a diagnosis when we feel pain or suffer incapacity; we expect the world of medicine to at once categorize, explain, and predict.

But today's diagnostic categories are not always sharp-edged and unambiguous, even if they exist as entities in accepted disease taxonomies. Emotional states present a particularly difficult problem. Like fevers, loose bowels, and seizures, anxiety, sadness, phobias, and compulsions have always been with us. But when are extreme feelings and atypical behavior considered sickness? Fear and sadness, for example, are both unavoidable—normal—aspects of the human condition, yet both can become the core of powerful emotions that bring pain, stigma, and social incapacity. When do such felt emotions and the behaviors they structure transcend the randomness of human idiosyncrasy and the particularity of circumstance and become what clinicians have over time become willing to recognize as disease? It is an elusive and ever-more-ubiquitous question.

Is depression a discrete thing or an agreed-upon set of behavioral responses and thresholds? Once diagnosed, it becomes a bureaucratic entity, a circumstantial variable helping structure interactions among sufferers, physicians, families, and—at the moment—insurers and pharmaceutical manufacturers. But though a real social entity, depression rests on shaky and contested epistemological grounds. Can a prolonged depression be an understandable response to circumstance—a kind of emotional reality testing—or is it always a disease process? How is one to differentiate the idiosyncratic and statistically predictable variation among people from the pathological and dysfunctional? How to tell the personal adjustment from the serious ailment?

Anxiety is equally and symmetrically elusive. Allan V. Horwitz has, in the present study, traced the complex and ambiguous but continuous history of the cluster of emotions and feelings we have come to call anxiety disorders. Humankind has always known fear, and evolution has presumably made it a normal—adaptive—component of humans as we perceive and anticipate our surroundings. But when do generalized anxieties, phobias, and compulsions become diseases? When does inappropriate fear cease to be a quirk of personality and become a clinically visible and legitimately treatable thing? Since Classical antiquity, physicians have been aware that fear could incapacitate, make "cowards" of soldiers, unhinge the ordinarily rational. That spectrum from normal to incapacitating anxieties has always been understood as an unavoidable aspect of the collective human condition. Since the mid-nineteenth century, however, as Western medicine has sought to understand disease in terms of discrete entities, anxiety has increasingly been regarded as the unstructured raw material from which to construct clinically usable disease categories. Phobias, compulsions, and anxiety states have been described, "discovered," and explained, from George Beard's neurasthenia in Gilded Age America and Sigmund Freud's anxiety neurosis in late-nineteenth-century Vienna, to the American Psychiatric Association's *DSM* (*Diagnostic and Statistical Manual*) categories in late-twentieth-century America. Though these categories are still unsettled and continuously being renegotiated, the drive to make anxiety a neatly disaggregated list of nameable things seems relentless. Anxiety disorders occupied fifteen pages in the APA's 1980 *DSM III*, eighteen pages in the revised edition of 1987, and fifty-one pages in the *DSM IV* of 1994. The fifth edition, of 2013, devotes ninety-nine pages to what had been considered anxiety disorders, though it defines separate categories for obsessive-compulsive and related disorders and trauma- and stressor-related disorders. The elaboration proceeds. Allan Horwitz has described an evolution that is both a substantive part of the modern experience and a microcosm symbolic of the way in which we have created conceptual and bureaucratic entities that—for better or worse—define us, constrain us, help us to think about who we are.

Charles E. Rosenberg

ACKNOWLEDGMENTS

The first known representations of anxiety are found in cave paintings from the Paleolithic era. They vividly depict sources of fear—usually dangerous predators such as lions, wolves, and bears—among our primeval ancestors. Sophisticated discussions of the nature and sources of anxiety emerged during the fourth century BCE in Hippocratic medicine and Aristotelian philosophy. The vast array of available works spans both a temporal range of thousands of years and a variety of medical, philosophical, religious, psychological, and sociological discourses. A book of this length can hardly encompass the huge sweep of material about anxiety. Therefore, I have limited the scope of the book to Western discussions and neglected the diverse understandings of anxiety found in, among other places, China, Japan, India, and the Middle East. In addition, the enormity of the range of literature has meant that I have made unusually heavy reliance on the writings of other historically oriented scholars. I have found especially valuable the studies of Germain Berrios, Joanna Bourke, Gerrit Glas, David Herzberg, Stanley Jackson, Michael MacDonald, George Makari, Mark Micale, Janet Oppenheim, Roy Porter, Charles Rosenberg, Andrew Scull, Edward Shorter, Andrea Tone, and Yi-fu Tuan. None of these outstanding researchers, of course, is responsible for the uses to which I have put their work.

I am grateful for the institutional support I received while writing this book. It was completed while I was a fellow at the Center for Advanced Study in Behavioral Science at Stanford University. I greatly appreciate the opportunity that the center and its director, first Stephen Kosslyn and then Iris Litt, gave me to spend a year in its unparalleled environment. As always, David Mechanic, the director of the Institute for Health, Health Care Policy, and

Aging Research at Rutgers University, provided invaluable support for my work. David is also responsible for creating at Rutgers a unique climate in which historical vision is seen as an essential framework for understanding current issues regarding health and health policy. My interactions at the institute with such gifted historians as Gerald Grob, Keith Wailoo, Elizabeth Lunbeck, and Nancy Tomes have profoundly shaped my own scholarship. I'm also grateful to Melissa Lane, Helene Pott, and Jamie Walkup for their ideas about improving this manuscript. My editor at the Johns Hopkins University Press, Jacqueline Wehmuller, provided insightful suggestions that have greatly enhanced the book, and Anne Whitmore at the Press was an unusually capable copy editor. I am especially appreciative of the efforts of distinguished historian Charles Rosenberg, the editor of the series of biographies of diseases in which this book appears. Charles provided astute advice from the book's initial conception to its final drafts. I thank him for the opportunity to extend my writing from its native sociological grounds to new historical territory.

ANXIETY

CHAPTER ONE

Afraid

Scientific views in the twenty-first century attempt to root the psychological and the physiological aspects of anxiousness in biological processes that include brain circuitry, neurochemicals, and genes. Anxiety and its disorders involve brain regions that are devoted to fear recognition, including the amygdala, prefrontal cortex, and hippocampus, and neurochemicals, such as GABA, epinephrine, dopamine, and serotonin. Neuroimaging techniques allow scientists to see how these neural networks activate in response to threats and let them pinpoint how different kinds of brains react to fearful stimuli. The mapping of the human genome has opened the pathway to examining the genetic correlates of anxiousness. Anti-anxiety drugs can now target the particular neurotransmitters related to this emotion.[1]

How do neuroscientists know whether what they are seeing represents a pathological or a natural form of anxiety? Modern societies have granted the preeminent power to define dysfunctional anxiety (and other mental disorders) to the psychiatric profession. Psychiatry's view of anxiety is embodied in the *Diagnostic and Statistical Manual* (*DSM*).[2] Since its landmark third edition, which was published in 1980, the *DSM* has used a medical model in which constellations of manifest symptoms define each disor-

der. It defines anxiety as a syndrome that occurs independently
of any unique qualities of the patients who carry its symptoms
or of the period of human history when it appeared. The *DSM*
assigns pathological anxiety conditions to a variety of categories:
panic disorder (with or without agoraphobia), agoraphobia without panic, specific phobias, social phobias, obsessive-compulsive
disorders, post-traumatic stress disorders, acute stress disorders,
and generalized anxiety disorders and several others. These conditions are presumably distinct from one another and from other
categories of mental disorder, such as depression and somatoform
disorders. The *DSM* definitions also use the number and severity
of symptoms to make sharp distinctions between people who have
disorders and those who don't.

In the United States (and, increasingly, worldwide), the *DSM*
provides the official definitions of anxiety disorders used by all
types of medical and mental health professionals. For the past
thirty years, its discrete diagnoses have been at the heart of the
entire enterprise of defining, studying, and responding to anxiety.
They are the basis of mental health practice, insurance reimbursement, drug prescription, public health statistics, government policy, legal liability, and cultural definitions of anxiety. The *DSM*'s
specific diagnoses have also come to underpin the large scientific
enterprise devoted to examining anxiety. Researchers attempt to
find the causes of and treatments for the various anxiety conditions, neuroscientists to isolate the particular brain circuitry that
underlies each, and epidemiologists to measure how often these
entities arise.

The most common current treatments for anxiety, which use
drugs that target presumably malfunctioning neurotransmitters,
also reflect a biomedical model. In 2008 American physicians
wrote more than 50 million prescriptions for specifically anti-anxiety medications and more than 150 million for antidepressants,
many of which were used for anxiety-related conditions. Moreover, usage rates of drug treatments are growing exponentially
while those of nonmedical therapies are declining.[3]

In certain respects, current knowledge about anxiety far exceeds that of previous eras. Imaging techniques can locate the regions of the brain where anxiety occurs with far greater precision than ever before. In addition, knowledge of the various neurochemicals involved in activating anxiety, which began to emerge only in the 1960s, allows new drugs to target the exact receptors related to anxiety. Likewise, geneticists recently became able to measure which alleles are associated with greater likelihoods of developing anxiousness. The brain structures and functions that these new technologies reveal are essentially timeless; they have operated in the same ways in humans at all times and in all places. If earlier students of anxiety had had the tools modern researchers possess, they presumably would have seen identical brain images among, say, ancient Greek soldiers facing battle, medieval peasants confronting an impending plague, and mid-twentieth-century psychoanalytic patients agonizing over their inadequate sexual performances. If previous assessments simply lacked techniques like brain imaging and gene sequencing that would have allowed them to accurately define and measure anxiety, then historical accounts would provide anachronistic characterizations of an emotion whose true description awaited the availability of more advanced technologies and measurements. That, however, is not the case.

SYMPTOM POOLS

"We must somehow," historian Edward Shorter observes, "draw upon the cultural symptom pool for models of illness to help us amplify and make sense of our own dim physical perceptions. Otherwise the mind cannot understand what the body is saying."[4] While current neuroimaging technologies uncover with remarkable precision the places in the brain where anxiety is located, they cannot reveal what anxious people are feeling or the ways in which they express their feelings. What people feel and express does not simply reflect physical sensations. The brain-based aspects of anxiety become manifest only through social understandings that shape how individuals define, display, and respond to physiological sensations. Current views of anxiety and its disorders, no less than those in the past, are infused with cultural templates, social influences, and material interests.

The *DSM* definitions supposedly reflect enduring and universal entities. Indeed, the manual relegates what it calls "culture-bound syndromes" to an appendix. Yet, the manifest symptoms its criteria use to define each disorder inevitably reflect culturally defined symptom pools. For example, in 1871 an army surgeon, Jacob Da Costa, coined the term "irritable heart" syndrome to describe the palpitations, cardiac pain, numbness of the arm, rapid pulse, shortness of breath, and sweating he observed among combat veterans of the American Civil War but could not relate to any physiological abnormalities in their hearts. Later, victims of "shell shock" during World War I commonly developed symptoms such as paralysis of limbs, inability to speak, or blindness, despite having no apparent somatic pathology. Vietnam veterans who suffered from what was named post-traumatic stress disorder often reported flashbacks to traumatic combat-related experiences, expressions that were very rare among combatants in previous wars. While common brain circuitry might underlie these diverse conditions, trauma survivors use whatever symptoms their cultural templates make available to them to express their anxiety. The psychological reexperiences of traumas that the current *DSM* emphasizes are not an abiding feature of a disease but a culturally specific expression.[5]

Cultural symptom pools are particularly important in describing anxiety conditions, which can assume a great variety of manifestations. Sometimes anxiety is not connected to anything specific but is more diffuse and general. At other times anxiety emerges in response to some definite object or situation, in which case it is generally called fear. Fear, then, is anxiety that is attached to a particular thing or circumstance. I use the term "anxiety" for both feelings that are amorphous and those that are directed at some definite threat. Therefore, as shorthand, I often use "anxiety" to refer to fear as well as to more generalized anxiety.

Psychiatrist John Nemiah's 1985 description captures the common experience of anxiety: "Mental and bodily functions find in anxiety a meeting place that is unparalleled in other aspects of human life."[6] Anxiety is often exhibited through disturbances in cardiovascular, gastrointestinal, and/or musculoskeletal systems. Indeed, the term "anxiety" derives from an Indo-Germanic root, *Angh*, which means narrowing, constricting, and tightening feelings, usually in the chest or throat.[7] In such cases, people might not be consciously aware of their anxiousness, although it is being expressed through stomach aches, heart palpitations, difficulties in breathing, and the like. Anxiety also has psychological aspects, ranging from unease and concern, through worry and dismay, to alarm, dread, fright, and terror. "In the daytime," Robert Burton wrote in 1621 about the concerns of anxious people, "they are affrighted still by some terrible object, and torn in pieces with suspicion, fear, sorrow, discontents, cares, shame, anguish, etc., as so many wild horses, that they cannot be quiet an hour, a minute of the time."[8] Such psychological experiences are difficult to reduce to physiological brain processes. In addition, anxiety can become manifest through behaviors, such as avoidance of fearful situations.[9]

Expressions of anxiety are not limited to somatic, psychological, and behavioral symptoms; they can be existential. Anxiety can be experienced as concern regarding one's place in the world or universe and may involve questions about the meaning of existence, the presence of God, and the possibility of an afterlife. "All existence," Sören Kierkegaard memorably stated, "makes me anxious, from the smallest fly to the mysteries of the Incarnation; the whole thing is inexplicable to me, I myself most of all; to me all existence is infected, I myself most of all. My distress is enormous, boundless; no one knows it except God in heaven, and he will not console me; no one can console me except God in heaven, and he will not take compassion on me."[10] Before the industrial era, such spiritual conceptions of anxiety were far more common than medical ones.

By its nature, anxiousness is a chameleonlike force that exhibits itself through a great variety of physical, mental, spiritual, behavioral, and other manifestations. Cultures shape these amorphous expressions into forms that both healing professionals and laypersons interpret as legitimate or illegitimate. Conditions that are widespread and well-recognized in one era, such as hysterical paralyses of limbs and fainting spells, disappear and reappear as another era's panic attacks and social anxiety.[11] This fluidity often confounds the attempts at precise delineation of specific conditions that modern medicine and psychiatry demand.

CATEGORIZATIONS

Medical and psychiatric history before 1980 generally depicted anxiety as a broad, nonspecific, and undifferentiated condition that was usually connected with other syndromes, especially depression. The large variety of distinct and categorical definitions of anxiety that arose in the *DSM-III* in 1980 is historically anomalous. By way of comparison, consider a recent description of an anxious person, which the novelist Paul Auster wrote about his mother:

> At the other end, the extreme end of who she was, there was the frightened and debilitated neurotic, the helpless creature prey to blistering assaults of anxiety, the phobic whose incapacities grew as the years advanced—from an early fear of heights to a metastatic flowering of multiple forms of paralysis: afraid of escalators, afraid of airplanes, afraid of elevators, afraid to drive a car, afraid of going near windows on upper floors of buildings, afraid to be alone, afraid of open spaces, afraid to walk anywhere (she felt she would lose her balance or pass out), and an ever-present hypochondria that gradually reached the most exalted summits of dread. In other words: afraid to die, which in the end is probably no different than saying: afraid to live.[12]

Although the terms Auster uses to describe his mother might fit into the tidy boxes of the *DSM* diagnoses of specific phobias, social phobias, agoraphobia, generalized anxiety, etc., her conditions are not distinct. Instead, they seem to express an underlying anxious temperament more than discrete yet co-occurring conditions. Auster's mother was an anxious *person* who was inseparable from her various fears. Likewise, conditions outside of the anxiety category, such as somatoform and mood disorders, accompanied her fears. The categorical entities of *DSM-III* have little resonance with this description, which the more generalized conceptions of anxiety found in most historical accounts of anxiety would have more accurately captured.

Indeed, biomedical research indicates that anxiety more closely conforms to the relatively undifferentiated and continuous conceptions that marked most of its history.[13] Neuroscientific, genetic, clinical, and epidemiological studies all fail to support the distinctiveness embodied in the categorical conceptions of the current diagnostic manual. Someone who has, for example, a social phobia is also highly likely to suffer from additional anxiety conditions. Common etiologies, prognoses, and treatments cut across the various anxiety diagnoses. Moreover, the various anxiety conditions share similar genetic vulnerabilities, risk factors, and treatments with depression: anxious people are more likely than not to also be depressed, just as most depressed people also suffer from anxiety. The *DSM*'s isolation of various specifically anxious conditions does not seem to reflect any underlying natural reality. The *DSM*'s deficiencies in defining mental disorders are so glaring that the National Institute of Mental Health (NIMH), the major federal agency concerned with research about mental illness, is considering abandoning the *DSM* classifications altogether. The NIMH is currently developing an alternative manual, the Research Domain Criteria (RDoC), which will attempt to use brain circuitry and genetic findings as the basis for a new diagnostic system that might supplement or even replace the *DSM*.[14]

If the *DSM* provides such inadequate definitions of anxiety (and other disorders), what accounts for its extraordinary success? The manual's disease classifications both reflect and reinforce the social framework in which the contemporary mental health professions operate. The entire professional, economic, and cultural foundation of the modern mental health enterprise rests on practitioners' ability to make specific diagnoses of diseases that are neither vague syndromes nor ordinary distress. The specificity of the *DSM*'s categories secured psychiatry's legitimacy as a medical discipline, because professionally credible medical specialties now require well-defined conditions. In addition, clinicians are reimbursed for treating only the kinds of categorical states that the *DSM* describes. Likewise, pharmaceutical companies can promote their products as remedies for only specific anxiety disorders. Mental health policy makers, too, gain credibility if they are viewed as responding to disease conditions as opposed to psychosocial problems. Specific diagnoses assure consumers of drugs and other therapies that they have legitimate medical conditions deserving of treatment. Patients willingly seek *DSM* diagnoses, which provide them their tickets to receiving professional help. Their therapists, of course, are more than happy to oblige them.

The practical usefulness of the *DSM*'s categorical system to patients as well as to clinicians, researchers, psychiatric institutions, insurers, drug companies, policy makers, and mental health advocates accounts for its becoming thoroughly embedded in modern societies. Like previous classifications, current definitions of anxiety disorders are social constructions that reflect a particular historical context far more than some underlying natural reality.

EXPLANATIONS

There is no doubt that anxiety is in some senses based in the brain, yet understanding anxiety as a brain-mediated process does not explain why the neurological circuitry related to fear recognition activates. In some cases, brain activity associated with anxiety can reflect biological factors such as abnormal levels of neurotransmitters or brain dysfunctions; but these same processes can also result from psychological mechanisms, cognitive interpretations, or personality vulnerabilities. Moreover, anxiousness often stems from dangerous and uncertain social environments.

Different cultures have used various medical, religious, philosophical, psychological, moral, and social frameworks to characterize anxiety and its disorders. The current emphasis on grounding anxiousness in neurochemistry in many respects echoes the dominant somatic paradigm of nineteenth-century psychiatry that explored how specific biological malfunctions in the nervous system make certain individuals vulnerable to anxiousness. "The brain is exclusively the organ of the mind," one of the most prominent phrenologists, Joseph Gall, declared in 1810; "the brain is composed of as many particular and independent organs as there are fundamental powers of the mind."[15] Well before Gall, a Hippocratic text noted: "Men ought to know that from the brain, and from the brain only, arise our pleasures, joys, laughter, and jests, as well as our sorrows, pains, griefs and tears . . . . It is the same thing which makes us mad or delirious, inspires us with dread and fear, whether by night or by day, brings sleeplessness, inopportune mistakes, aimless anxieties, absentmindedness, and acts that are contrary to habit."[16] Recent psychiatric studies that emphasize the particular neural networks and neurochemicals associated with anxiety represent a resurgence of this somatic view.

Another major tradition focuses on the psychological vulnerabilities and personality styles that lead to nervous conditions. For example, Aristotle (384–322 BCE) distinguished anxious personalities from anxiety disorders: "The man who is by nature apt to fear everything, even the squeak of a mouse, is cowardly with a brutish cowardice, while the man who feared a weasel did so in consequence of disease."[17] Men who feared everything (cowards) had anxious personalities, which were abiding predispositions that led them to be prone to constant states of anxiousness. Such anxious temperaments, albeit undesirable in Aristotle's view, were not disordered. In contrast, fears of particular objects that were not rational sources of fear (e.g., weasels) likely resulted from some disease.

Many people's temperaments lead them to be prone to constant states of anxiousness. Robert Burton's 1621 compendium of 2000 years of thinking about melancholy focuses on such dispositions, and observes that the "melancholic . . . upon every small occasion of misconceived abuse, injury, grief, disgrace, loss, cross, rumour, etc. yields so far to passion, that his complexion is altered, his digestion hindered, his sleep gone, his spirits obscured and his heart heavy . . . and he himself overcome with melancholy."[18] A popular modern blog about anxiety is full of stories of such anxious personalities. One, for instance, notes:

> Like many people, I like to set aside a few hours every day, generally between 3 and 6 a.m., to lie quietly thinking about everything that could go horribly wrong with my life and all the ways in which I am negligent and reprehensible. I have spasms of panic over things I shouldn't have written, or, worse, things I should have; I regret having spent all the money and wonder where more money might ever conceivably come from; I wish I'd kissed girls I didn't, as long ago as 1985. I'm suddenly convulsed with remorse over mean things I did in middle school . . . I force myself to choose my least favorite death (drowning).[19]

These anxious dispositions can originate at birth and persist throughout a person's lifetime.[20]

A third tradition, initiated in Scottish physician George Cheyne's *The English Malady* (1733), highlights societywide factors such as stressful life conditions, dissolute patterns of living, rapid social changes, economic insecurities, and unfulfilled desires that are conducive to anxiety. Well before Cheyne wrote, patients themselves were reporting anxieties caused by such factors. For example, the principal sources of anxiety among the patients of a seventeenth-century English physician, Richard Napier, were worries over love, marriage, death, and, especially, money. "The traditional hazards of economic life caused a great deal of reported distress among Napier's patients . . . . Debt was by far the greatest single source of anxiety." Recent research indicates that socially disadvantaged groups are still especially likely to suffer from anxiety. One study, for example, concludes that "anxiety in poor mothers is usually not psychiatric, but a reaction to severe environmental deficits." An emphasis on the environmental roots of anxiousness also persists among behavioral psychologists, who highlight how external rewards and punishments lead to the emergence and maintenance of anxious behaviors, and among sociologists of mental health, who focus on stressful life events and social roles that are conducive to distress.[21]

Factors related to biology, psychology, life history, and social and natural environments are omnipresent potential causes of anxiety. Different societies emphasize whichever perspective best fits their general cultural, social, and material conditions. For example, the social model that framed conceptions of anxiety in the United States during the 1950s and 1960s was compatible with the more general cultural confidence in the power of social institutions to shape behavior that prevailed during that period. Likewise, the current primacy of neuroscientific views reflects the growth of belief systems emphasizing nature over nurture, which are compatible with the more individualistic worldviews that have marked recent decades.

Biological, psychological, and social emphases, as well as those that see anxiety as the result of moral flaws or spiritual imperfections, have woven in and out of characterizations at various times. In some periods, a single form of explanation has gained undisputed prominence; in others, diverse schools of thought have simultaneously competed for adherents. Moreover, a number of prominent models, from holistic Hippocratic medicine to the current biopsychosocial paradigm, have tried to integrate biological, temperamental, and environmental factors into a single system involving a general disturbance in the balance between individuals and their environments.[22] Prevailing models of anxiety have represented not just the state of scientific knowledge but also the social, cultural, economic, and professional conditions of the era.

RESPONSES

The current emphasis on using drugs to treat anxiousness is not new. Since Hippocratic times, healers have employed various substances, especially alcohol and the opiates, as frontline responses to anxiety. Indeed, the famed eleventh-century Persian physician Avicenna encouraged melancholics "not to drink only, but now and then to be drunk."[23] Before the twentieth century, folk healers, herbalists, and apothecaries, as well as physicians, regularly employed drugs as part of their arsenal to allay anxiety. Until recently, however, drugs were often part of more generalized healing regimes that focused on changing lifestyle, sleep patterns, diet, and exercise. "All [medical] practice," historian Charles Rosenberg indicates of the period before the nineteenth century, "was holistic, and prevention and therapeutics were in some ways indistinguishable."[24]

Spiritual responses have provided common alternatives to medical views of anxiety. Although its importance has oscillated through history, religion has simultaneously created anxiety —over sinfulness, eternal life, and salvation—and provided the means of controlling anxiousness. "God is our refuge and strength, a very present help in trouble," asserts one of the psalms in the Bible, "therefore we will not fear, though the earth be removed, and though the mountains be carried into the midst of the sea; Though the waters thereof roar and be troubled, though the mountains shake with the swelling thereof."[25] Relief from anxiety can stem from believing in God's or religion's healing power. Consider the typical response to an eclipse of the sun in the medieval world recorded by a Franciscan monk:

> In the year of our Lord 1239 there was an eclipse of the sun, wherein the light of day was horribly and terribly darkened, and the stars appeared. And it seemed as though night had come, and all men and women had sore fear, and went about as if bereft of their wits, with great sorrow and trembling. And many, smitten with terror, came to confession, and made penitence for their sins, and those who were at discord made peace with each other.[26]

Magic has also been used to alleviate anxiety. Throughout history folk healers, sorcerers, and witch doctors have employed a variety of techniques to protect people from sources of danger and to relieve their anxious emotions.[27] While religious, magical, and other spiritual techniques remain popular responses to anxiety in the modern world, they have been overshadowed by drug treatments to an extreme degree. As one modern commentator on Kierke-gaard asks, "Is there any doubt that were he alive today he would be supplied with a refillable prescription for Xanax?"[28]

Philosophers and, more recently, psychologists have developed cognitive and behavioral regimes to overcome anxiety. For example, the basis of Stoicism, the dominant school of Western philosophy for five centuries, was to train individuals to cultivate inner indifference to external sources of anxiety. Psychologists, as well, use therapies that encourage people to visualize their fears and perceive them as finite and manageable.[29]

Biological, religious, magical, and cognitive means of controlling anxiety have waxed and waned in prevalence throughout history. Which healing regimes gain prominence at any particular time depends on an array of factors that include cultural ideologies, the availability and cost of treatments, the degree of stigma attached to different responses, and practitioners' ability to establish professional dominance. While considerable evidence shows that current anti-anxiety medications are effective, whether they produce better results than did earlier substances, cognitive systems, or holistic treatments is an unresolvable question.

NORMALITY AND ABNORMALITY

Cultures provide not only symptom pools and classifications to channel expressions of anxiety, models for credible explanations, and treatments of anxiety but also definitions of when anxiety is natural and when pathological. Societies that have faced the most severe and chronic anxiety-producing conditions have been the least likely to define anxiety as disordered. Before the modern age, widespread violence, famine, disease epidemics, precarious material circumstances, and the like made anxiety an omnipresent part of existence. Anxiety only stood out and began to be widely defined and treated as a disorder during the nineteenth century, once people could expect a basic level of safety and certainty in their lives.[30]

Identical symptoms and their resultant neural indicators can signify natural responses at some times and anxiety disorders at others. Throughout history, commentators have recognized that anxiety expectably arises in threatening and uncertain situations. Aristotle distinguished normal from abnormal fears and anxieties, pointing out that the former arise in dangerous situations and so are contextually appropriate while the latter do not. Natural fears, according to him, "may be felt both too much and too little, and in both cases not well; but to feel them at the right times, with reference to the right objects, towards the right people, with the right motive, and in the right way is what is both intermediate and best, and this is characteristic of virtue."[31] Unnatural fears and anxieties, in contrast, were either excessive or deficient relative to the degree of danger inherent in a situation.

Conversely, nonthreatening environments have always been defined as being free of fear. A Sumerian cuneiform text from 4000 BCE illustrates such a golden age, one without dangers:

> Once upon a time, there was no snake, there was no scorpion,
> There was no hyena, there was no lion,
> There was no wild dog, no wolf,
> There was no fear, no terror,
> Man had no rival.[32]

Anxiety-free environments usually lie in the distant past, not the present.

Observers have always recognized the existence of pathological forms of anxiety, although the definitions have varied over time. "Neurotic anxiety is anxiety in regard to a danger which we do not know," wrote Freud, speaking of people who are anxious despite having nothing to fear.[33] Fears that people associate with objects or situations that aren't actually dangerous, such as innocuous animals, shopping in a supermarket, or attending a public lecture, can also be viewed as signs of disorder.[34] Sometimes, the severity of anxiousness can be so overwhelming that it leads to a breakdown of functioning: "Many men are so amazed and astonished with fear," English vicar Robert Burton observed in the seventeenth century, that "they know not where they are, what they say, what they do; and (that which is worse) it tortures them, many dayes before, with continual affrights and suspicion."[35] Inversely, people who are unable to become anxious in very dangerous situations are considered to have anxiety disorders—or a lack of sense. The Dutch Renaissance humanist Erasmus wrote about his flight from a plague-ridden area, "Really, I consider total absence of fear in situations such as mine, to be the mark not of a valiant fellow but of a dolt."[36]

No natural dividing line marks where normal anxiety ends and pathological conditions begin. Contextual considerations and cultural beliefs, which are intrinsically historically specific, are unavoidable aspects in separating appropriate from inappropriate types of anxiety. For example, an intense fear of witches, which would have been natural in sixteenth-century English villages, would be unrealistic and a possible sign of disorder in modern societies. Similarly, the Quechua Indians of Peru can develop anxiety conditions resembling PTSD after seeing old Incan ruins, hardly a likely response outside their culture.[37] A recurrent issue in the history of anxiety has been how to disentangle normal and pathological anxiousness.

In contrast to most of history, the social context of modern mental health practice insures that the current divisions between normal and pathological anxiety minimize the former and maximize the latter. Many influential groups benefit when conditions are defined as "mental disorders" as opposed to normal anxiety. Mental health professionals are reimbursed for treating anxiety disorders; drug companies must advertise their products as remedies for disordered conditions; researchers receive funding for studying anxiety disorders; policy makers are on safe political ground when they target illnesses; and mental health advocates gain legitimacy for obtaining services for the disordered. Moreover, patients can only get treatment and release from blame when their symptoms are viewed as disorders. None of these benefits accrue from calling anxiety "normal."

The symptom-based *DSM* diagnoses might have little scientific validity but they do have the advantage of labeling all anxious people who meet their criteria for any reason as having "disor-

ders." The *DSM* definitions encompass people whose symptoms
arise from threatening or uncertain circumstances, their temperaments, or evolutionary design, as well as from brain dysfunctions.
Therefore, their application leads to extraordinarily high estimates
of anxiety disorders—nearly a third of Americans meet *DSM* criteria for at least one anxiety disorder diagnosis at some point in
their life.[38] Many groups profit from the huge pools of potential
patients, consumers, and victims that result from the manual's
classifications of the anxiety disorders. Such social considerations
inevitably guide decisions about where natural anxiety ends and
anxiety disorders begin. This is as true now as in the past.

STUDYING ANXIETY THROUGH HISTORY

What we know about anxiety over the course of history stems
largely from medical, philosophical, theological, and literary
depictions of the condition. For the most part, then, this book
surveys conceptions of anxiety among physicians, psychiatrists,
philosophers, and theologians. It is difficult to know how closely
these sources conformed to and influenced how ordinary people
experienced anxiety. For example, physicians, not patients, write
medical histories, so the stories they tell might more accurately
reflect their conceptions than those of their clients.[39] In addition,
the farther back in time one moves, the more likely it is that extant
documents report only the most florid and visible cases of anxiety.
Before the expansion of outpatient treatment of anxiety at the end
of the nineteenth century, the condition rarely rose to a level of
social threat that made medical commentators or public authorities take written note of it. The ways common people experienced
and coped with anxious emotions have largely been lost to history.

Another dilemma regards how to establish comparability in
emotional states across widely varying times and places. On the
one hand, there is little doubt that anxiety is a universal, biologically grounded phenomenon. Physiological changes in heart rate,
blood pressure, adrenaline, breathing, and neural activity seem
relatively constant across cultures. Members of different cultures
also show fairly consistent facial expressions of fear, such as raised

eyebrows, darting eyes, and tense lower eyelids.[40] Although contemporary students of emotions rarely agree on the presence of common emotions across human societies, nearly all recognize fear and anxiety as among a small number of primary emotions. Despite the arguments of some social constructionists that emotions are linguistic constructions, humans would experience fear even if they had never heard others speak about it.[41] While people in diverse cultures and historical periods fear different things and express their fears in different ways, in all settings fear arises when people believe they are endangered. Anxiety cannot be reduced to cultural definitions.

On the other hand, the powerful cultural influences on how anxiety is expressed create the problem of how to tell if a uniform condition underlies different historical manifestations. If cultural norms affect the concepts and terms that define symptoms as well as the overt appearances of anxiety, how is it possible to know if the same phenomenon is being studied in different time periods? For example, when symptom pools channel anxiety into physical symptoms in one era but into psychic symptoms in another, what means can show that some underlying feature makes the two styles of symptom expression instances of the same state? Any history of anxiety faces the quandary of how to study a phenomenon that is at one and the same time a basic biological process and a deeply culturally infused condition.

CONCLUSION

In certain respects, current knowledge about anxiety is more advanced than at any point in the past. The abilities to map the brain regions where fear recognition occurs, to measure neural activity connected with anxiety, and to locate the genetic alleles that create vulnerabilities for developing anxious conditions are undoubtedly major and recent scientific achievements. Knowing how to paint a far better neurological picture than ever before, however, does not resolve many of the most important issues in understanding anxiety and its disorders. We still lack knowledge of how brain images are associated with subjective experiences, represent natural or pathological anxiety, or correspond to varying cultural templates. The great paradox of anxiety in the twenty-first century is that, despite unprecedented progress in capacities to view neural networks, neurotransmitters, and genes, our definitions of anxiety disorders, understandings of what causes them, and treatments for them might not be significantly better than those possessed by Hippocratic medicine.

No less than in previous eras, current conceptions of anxiety and its disorders reflect the social matrix in which they arise, are identified, and are treated. The history of anxiety and its disorders reflects an ongoing struggle over how to assimilate the amorphous variety of forms that anxiety can assume into taxonomies that inevitably reflect prevailing cultural and social influences as well as scientific knowledge. The transformations and commonalities of anxiety through the ages that the following chapters describe show how biological properties and cultural interpretations intrinsically interact and mutually influence each other.

CHAPTER TWO

Classical Anxiety

THE HIPPOCRATIC CORPUS

The roots of current conceptions of anxiety and its disorders date back to the emergence of Classical Greek civilization in the fifth and fourth centuries BCE. Medical and philosophical writings during this period reoriented thinking about human behavior away from mythology and religion and provided an empirical and observational foundation for the study of health and disease. During the Classical period, physicians and philosophers produced the first recorded, sustained discussions about sanity and madness, including normal and pathological forms of fear and anxiety. The emergence of Hippocratic medicine in the fifth century BCE provided a naturalistic foundation for the study of mental illness that, as channeled through later Galenic medical culture, provided a template for studying anxiety and other mental conditions that persisted in medicine for thousands of subsequent years.[1]

At the same time, conceptions of anxiety were deeply rooted in the particular values of Classical civilizations. While medical interventions were sometimes necessary, anxiety was more likely to be attached to valued and disvalued character traits than to organic dysfunctions in need of somatic correction. Anxiety was thoroughly embedded in moral discourse. Men (the Greeks and Romans rarely considered women) were characterized as brave or cowardly depending on whether they controlled or succumbed to their anxious emotions. If Classical conceptions in certain respects continue to constitute the core intuitions about the inherent nature of anxiety, they also illustrate the close connection of notions of anxiety with the values that prevail in a particular time and place.

THE NATURE OF ANXIETY

Classical discussions were more likely to focus on normal expressions of fear and anxiety than anxiety disorders. Greek culture was intensely concerned with the maintenance of social esteem. The culture's essential values included demonstrating courage through willingness to fight, bravery in battle, and steadfastness during confrontations with enemies. Conversely, cowardice—seen as the inability to remain resolute in threatening situations—was a stigma to be avoided at all costs. Inappropriate display of fear in battle was the most shameful possible behavior for a Greek male and could lead to ridicule, contempt, and unbearable loss of face.[2]

These cultural values profoundly shaped Greek (and, later, Roman) conceptions of anxiety. The Greeks defined anxiety broadly as "the expectation of evil," which included "all evils, e.g. disgrace, poverty, disease, friendlessness, death."[3] Nevertheless, they focused on fear that arose in dangerous situations, in particular, combat. Aristotle (384–322 BCE) provided the most extensive discussion, emphasizing how normal fear and anxiety stem from threatening situations: "*Let fear, then, be a kind of pain or disturbance resulting from the imagination of impending danger, either destructive or painful* . . . . For that is what *danger* is—*the proximity of the frightening.*"[4] This definition was eminently contextual: symptoms characteristic of anxiety naturally arise when people perceive some impending danger.

All men naturally feared battle, they declared. "Hence," the Roman philosopher Seneca (4 BCE–65 CE) noted, "even the man who is generally very brave grows pale when putting on his armor; the fiercest of soldiers is weak at the knees when the signal to engage is given; a great general's heart pounds before the lines of battle meet."[5] Even the bravest men felt fearful; what was essential for the maintenance of social esteem was their ability to control their fears. The essential difference lay in how someone dealt with his natural propensity to be afraid. This distinction was more a moral than a physiological or psychological one.

Consider the Iliad's comparison of how cowardly and brave men face battle:

> The skin of the coward changes color all the time,
> he can't get a grip on himself, he can't sit still,
> he squats and rocks, shifting his weight from foot to foot,
> his heart racing, pounding inside the fellow's ribs,
> his teeth chattering—he dreads some grisly death.[6]

Fright, then as now, was displayed through bodily changes such as blushing, fidgeting, pounding heart, and chattering teeth.[7] Someone who openly displayed such symptoms in combat was a coward and thus faced disgrace. Herodotus, for example, noted that Aristodemus was dishonored for missing the battle of Thermopylae because of eye trouble and received the epithet "he who trembled."[8]

The courageous response of the brave warrior contrasted with the coward's:

> But the skin of the brave soldier never blanches.
> He's all control. Tense but no great fear.[9]

The honorable man displayed his character and lived up to social standards by containing his anxious feelings. This intense emphasis on honor led Classical societies to place particular emphasis on the control of fear; fearfulness indicated a failure to live up to their most basic values. Anxiety was more likely to be cast in a framework that defined actions as courageous or cowardly than as a medical condition that physicians should treat.

Aristotle provided a philosophical grounding for valuing or disrespecting the ways men responded to danger. He defined as most worthy those whose behavior occupied the ground between the extremes of excessiveness and deficiency. Men who were fearless in the face of danger were rash just as those who feared everything were cowardly: "The coward, the rash man, and the brave man, then, are concerned with the same objects but are differently disposed toward them; for the first two exceed and fall short, while the third holds the middle, which is the right, position."[10] The appropriate response to danger was bravery: "The man, then, who faces and who fears the right things and from the right motive, in the right way and at the right time, and who feels confidence under the corresponding conditions, is brave; for the brave man feels and acts according to the merits of the case and in whatever way reason directs."[11] Brave men, unlike the fearless, felt fear but were able to control their fright.

The Greeks and the Romans embedded biologically grounded feelings in a cultural matrix. It was accepted that men naturally became anxious in dangerous situations, but the courage or cowardice of their response was defined in accordance with cultural value systems.[12]

SOCIAL ANXIETIES

The primary, but not exclusive, focus of Classical thinkers was how dangerous situations, such as a battle or the sinking of a ship, led to anxiety. They also examined how certain kinds of social relationships naturally led to anxiousness. Greek society featured strongly entrenched notions of social status that specified precise ways that men of different statuses ought to behave toward each other.[13] For instance, sons were totally subordinate to their fathers, holding many duties but no rights. Social anxieties, like responses to dangerous situations, were viewed in moral terms: positions in social hierarchies determined what emotions were appropriate or inappropriate for people to feel.

The Greek Stoic philosopher Epictetus (55–135 CE) illustrated how people's positions in status hierarchies determined what was normal for them to fear:

> Therefore Zeno, when he was to meet Antigonus, felt no anxiety. For over that which he prized Antigonus had no power; and those things over which [Antigonus] had power, Zeno did not regard. But Antigonus felt anxiety when he was to meet Zeno, and with reason, for he was desirous to please him; and this was external ambition.[14]

Men naturally feared the wrath of their social superiors, but not their inferiors. Because Zeno was a social superior of Antigonus, he should not fear his subordinate; conversely, Antigonus would appropriately become anxious when he anticipated meeting Zeno. The powerful typically displayed the opposite attribute of fear—confidence: "Men have confidence . . . when a thing should not be fearsome to one's equals, nor to one's inferiors nor to those to whom one believes one is superior . . . ."[15] Accordingly, people should not fear things that those beneath them did not fear. Fear and confidence were less attributes of individual personalities than characteristics of one's social status.

Certain types of social situations were also accepted sources of anxiety. The Roman philosopher Seneca indicated the natural propensity of all men to become anxious when anticipating speaking in public: "the most eloquent orator's scalp tightens as he prepares to speak."[16] Epictetus elaborated on the nature of this type of social anxiety:

> When I see anyone anxious, I say, what does this man want? Unless he wanted something or other not in his own power, how could he still be anxious? A musician, for instance, feels no anxiety while he is singing by himself; but when he appears upon the stage he does, even if his voice be ever so good, or he plays ever so well. For what he wishes is not only to sing well but likewise to gain applause. But this is not in his own power.[17]

Social anxiety naturally accompanied fears of social evaluation as people anticipated others' judgments of their performances.

In addition to dangerous situations and certain types of social relationships and situations, cultural beliefs, especially those associated with religion, could generate intense fear. Greek drama is full of characters who fear the anger of gods that they or their ancestors had knowingly or unknowingly offended. Religious anxieties could be particularly terrifying because they were vague and impossible to attach to any particular cause. Conversely, men were expected not to feel anxious when their relations with the gods were on good footing.[18] The Greeks also intensely feared moral pollution, which was especially anxiety-provoking because it could be both infectious and hereditary and so needn't result from one's own actions. Socrates, for example, noted "forms of misery which have appeared in certain families in consequence of the wrath of the gods which has been stirred against them in the past."[19] Because men could do nothing to avoid this sort of pollution, it was an especially powerful source of anxiety.

ANXIETY DISORDERS

Greek medicine was not diagnostically oriented and it recognized only a few basic types of mental disorder, most importantly, phrenitis (acute mental disturbance with fever), mania (acute mental disturbance without fever), and melancholia (chronic mental disturbance). Anxiety was an integral aspect of the broader condition of melancholia and was closely linked to depression: "when fear and sadness last for a long time, this is melancholy."[20]

Judging from the core medical texts surviving from the Classical period, medicine rarely viewed anxiety as a free-standing condition. A combination of anxious concerns and nameless fears, depressive symptoms such as blackness of mood and suicidal impulses, and paranoid tendencies such as sullen suspiciousness characterized melancholic conditions.[21] One Hippocratic text described a condition marked by "anxiety, restlessness, dread . . . fear, sadness . . . fretfulness . . . pains at the praecordium [front of the chest overlying the heart]."[22] Similarly, Galen (131–201 CE) indicated that, although "each [melancholic] patient acts quite differently than the others, all of them exhibit fear or despondency."[23] He went on to note, "Therefore, it seems correct that Hippocrates classified all their symptoms into two groups: fear and despondency."[24]

Although anxiety was seen to naturally arise in threatening contexts, anxiety in melancholic disorders violated commonsense assumptions about what were appropriate behaviors in given situations. A disorder was indicated if anxiety didn't appear when it should, did emerge in the wrong situations, or was of grossly excessive intensity given the milieu. Both deficient and excessive amounts of fear in a given context could be signs of disorder.

Aristotle observed, "Of those who go to excess he who exceeds in fearlessness has no name . . . , but he would be sort of a madman or insensitive to pain if he feared nothing, neither earthquakes nor the waves, as they say the Celts do not."[25] He thought the Celts disordered because they didn't fear things that he assumed any reasonable person ought to fear. Aristotle also distinguished fearlessness, which caused men to rush into the face of danger without regard to the consequences, from courage, which caused men not to succumb to the fear that they did experience. Fearlessness was not a sign of courage but an inappropriate response to manifestly dangerous situations.[26]

More commonly in Classical society, anxiety disorders were perceived as involving excessive amounts of fear. Often, undue anxiety indicated moral weakness, not medical disorder. "The man who exceeds in fear is a coward; for he fears both what he ought not and as he ought not," wrote Aristotle.[27] However, some men who became anxious in contexts that were not normally defined as dangerous could be viewed as having melancholic disorders. For example, the noted Roman physician Celsus (ca. 25 BCE–ca. 50 CE) described a type of melancholy characterized by "unmotivated fears, anxiety, and sadness."[28] The term "unmotivated" indicates that contextually appropriate, "motivated" fears would not signify disorders.

Aretaeus of Cappadocia (ca. 150–200 CE) observed similar cases of melancholy: "Sufferers are dull or stern: dejected or unreasonably torpid, without any manifest cause. . . . Unreasonable fears also seize them."[29] "Without any manifest cause" and "un-reasonably" distinguished the disorder melancholy from natural states of sorrow and fear. People with melancholia experienced the symptoms of fright, such as pounding heart and disturbances in the chest, although nothing alarming had occurred. In addi-tion, these symptoms did not result from false beliefs that danger was imminent, even though it wasn't; instead, the afflicted per-sons experienced intense fear despite *knowing* that they were not in danger. Therefore, their anxiousness appeared unreasonable to the sufferers themselves as well as to others. Unlike natural fears, which arose in dangerous contexts, melancholic fear emerged from some internal derangement rather than an appropriate ex-ternal cause. Such fears were not signs of cowardice, which was connected to inappropriate actions in dangerous situations, and so were more likely to be framed in terms of medical discourse.

Occasionally, dangerous situations were so fearsome that they led to serious frights that unhinged the mind (melancholy "with cause"). The Roman philosopher Lucretius (99–ca. 55 BCE) de-scribed the physical effects of extreme shocks:

> But when the mind is moved by shock more fierce,
> We mark the whole soul suffering all at once
> Along man's members: sweats and pallors spread
> Over the body, and the tongue is broken,
> And fails the voice away, and ring the ears,
> Mists blind the eyeballs, and the joints collapse,
> —Aye, men drop dead from terror of the mind.[30]

A fifth-century CE Roman physician, Caelius Aurelianus, de-scribed another case of a man who was so frightened by the ap-proach of a crocodile that he falsely imagined that the animal bit off his leg and hand. Here, a source of normal fear—the sudden appearance of an imminent threat—caused a delusional, disor-dered state.

Fears that involved delusionary ideas were considered disor-dered. Galen described this sort of fear:

As for instance, one patient believes that he has been turned into
a kind of snail and therefore runs away from everyone he meet
lest [its shell] should get crushed; . . . Again, another patient is
afraid that Atlas who supports the world will become tired and
throw it away and he and all of us will be crushed and pushed
together. And there are a thousand other imaginary ideas.[31]

The content of such delusions also illustrates how cultural templates influence displays of madness. In the Classical age these
included becoming frightened by hearing flute music, believing
that Atlas could crush one if he dropped the world, and fear of
shattering because one was a piece of pottery or glass.[32]

In addition to its intrinsic connection to melancholia, anxiety was a prominent symptom of other disorders, especially of
agitated people who were afflicted with mania. For example, in
the *Ion*, Plato describes a type of poetic possession and madness
in which the poet's eyes fill with tears, his hair stands on end, and
his heart pounds with fear.[33] Similarly, in the first century BCE,
empiricist physicians described the typical sequence of symptoms
in mania as: "cephalea (headaches), anxiety, mental confusion and
delirium."[34] Later, in the fourth century CE, the Roman clinician
Posidonius likewise observed of a typical manic patient: "The patient laughs, sings, dances . . . he bites himself . . . sometimes he is
wicked and kills . . . sometimes he is anxious and seized by terror
or hate . . . ."[35]

Occasional Hippocratic case histories depicted cases of anxiety disorders that were not connected to more general categories
such as melancholia or mania and were not prompted by rational
sources of anxiety. In the third century BCE, Andreas of Charystos described two anxiety neuroses, aerophobia, an intense fear of
open spaces, and pantophobia, the fear of everything.[36] Caelius
Aurelianus described the condition of hydrophobia, the fear of
water, as being characterized by "anxiety, fear, insomnia . . . tor-
por, stupor . . . tremor . . . stretching of the nerves . . . difficulty
in breathing."[37] He also described the irrational fear of Democles,

who could not walk alongside a precipice, pass over a bridge, or step over a ditch, even though he could walk within the ditch.[38] Another example of a specific anxiety disorder was a condition which "usually attacks abroad, if a person is travelling a lonely road somewhere and fear seizes him."[39] One more case concerned a young man who lost control and suffered from an unbearable fear of flute music when it was played at nighttime banquets, although he had no fear of this music when it was played during the day. Such behaviors stemmed from incomprehensible sources of fear and, therefore, were judged to be signs of insanity.

Few such cases are known in which anxiety was the primary symptom. One reason is that mental illnesses became a source of interest in ancient societies only when they disrupted civic order or jeopardized property. Medical jurisdiction over mental illness at the time was limited to the most blatant forms of psychological dysfunction, which rarely included anxious states that disrupted only the sufferer's life. Such conditions seldom rose to the level of threat or severity that would gain public notice and notoriety. Merely idiosyncratic deviations that harmed no one but the afflicted individual would rarely stir enough communal interest to arouse concern and be recorded. Most anxiousness that would now be defined through medical idioms was then framed in alternative vocabularies, such as idiosyncrasies of character or spiritual malaise. Anxiety in the ancient world was more likely to be seen as a sign of moral failure than as a medical dysfunction.

CAUSES OF ANXIETY

Hippocratic writings rarely focused on distinct causes of anxiety conditions. The foundational Hippocratic principle was that health was a state of equilibrium within the body while disease was a disturbance of this balance.[40] A variety of factors, including diet, lifestyle, living conditions, and atmospheric elements could lead to an imbalance. The humors were what needed to be kept in balance. The Greeks viewed mental disease, like disease in general, in terms of the four basic bodily humors: blood, phlegm, yellow bile, and black bile. Each humor possessed two of the four basic qualities of hot, cold, moist, and dry. Blood was associated with warm and moist qualities, phlegm with cold and moist, yellow bile with warm and dry, and black bile with cold and dry. When the humors were in balance with one another, a healthy state resulted. Diseases stemmed from too much or too little of one of these humors.[41]

Mental illnesses were connected to an excess of black bile. Galen described the consequences of a brain invaded by black bile, which could lead to a melancholic condition: "The humour, like darkness . . . invades the seat of the soul, where reason is situated . . . melancholy then arises. . . . As children who fear darkness, so adults become when they are the prey of the black bile, which supports fear . . . . For this reason the melancholics are afraid of death and wish for it at the same time . . . they avoid light and love darkness."[42] Yet, mental disturbances that resulted from an excess of black bile were not localized but disrupted the holistic relationship between individuals and their surroundings. The Hippocratics rejected any sharp dichotomy between internal and environmental forces. Moreover, while diseased bodies could lead to diseased minds, the converse was also the case: upset minds often resulted in bodily disturbances.

Classical thinkers also recognized that certain sorts of personalities were predisposed to be more anxious than other types. Ideal personalities contained a balance of the four humors. Excessive blood, phlegm, yellow bile, or black bile led to sanguineous, phlegmatic, bilious, or melancholic temperaments, respectively.[43] Melancholic temperaments in themselves were not a sign of disease, but possessing one made a person more likely than others to be fearful when confronting danger.

Cicero (106–43 BCE) distinguished relatively abiding anxious aspects of personalities from anxiety that developed in threatening and dangerous situations: "Trait anxiety (*anxietas*) differs from state anxiety (*angor*) in the sense that those who sometimes show fear are not necessarily always anxious, nor do those who are always anxious necessarily show fear in all situations."[44] *Anxietas* was a relatively stable predisposition while *angor* was more transitory and marked by fearful outbursts.⁴⁵ Aristotle discussed how people with different temperaments showed variable responses in the face of danger, and he mentioned the influence of the quality of the humor involved: "When some alarming news is brought, if it happens at a time when the mixture [of black bile] is cooler, it makes a man cowardly; for it has shown the way to fear, and fear has a chilling effect. Those who are terrified prove this; for they tremble. But if the bile is hot, fear reduces it to the normal and makes a man *self-controlled and unmoved*.⁴⁶ Aristotle thus acknowledged temperamental variability of the natural tendencies to be afraid in dangerous situations.

Galen likewise noted how various personality styles affected the likelihood of becoming afraid, observing that "external darkness renders almost all persons fearful, with the exception of a few naturally audacious ones or those who were specially trained."⁴⁷ He presumed that humans were naturally predisposed to fear darkness, although certain temperaments ("naturally audacious ones") or special training could overcome such normal fears. Both contextually related fearfulness and anxious personalities were associated with normal, not disordered, anxiety.

THE CONTROL OF ANXIETY

Greek and Roman responses to anxiety also reflected those societies' cultural values and institutional needs. Classical thinkers understood fear to be an especially powerful emotion that naturally emerged in response to danger. They also understood that it was very difficult for conscious decisions to override the power of fear. Therefore, only deeply rooted social norms could overcome the compelling force of anxious emotions. A central social concern in ancient Greece and Rome was to prevent natural fear and anxiety from arising in dangerous situations, especially combat.

Courage was the moral virtue related to controlling fear in the face of danger. It was not the same as fearlessness. Courageous responses were not natural but required long periods of disciplined training that gave one the ability to overcome natural passions and inculcated steadfastness in dangerous situations. Such education didn't eliminate fear but taught the courageous person how to properly apprehend dangerous situations, moderate fears, and make appropriate responses to them.[48]

Ideally, training to resist fear in dangerous situations began at an early age. According to Aristotle, "It makes no small difference, then, whether we form habits of one kind or another from our very youth; it makes a very great difference, or rather *all* the difference."[49] Such training developed character and allowed men to automatically respond to danger without giving way to fear:

> It was the mark of a brave man to face things as they are, and seem, terrible for a man, because it is noble to do so and disgraceful not to do so. Hence also it is thought the mark of a braver man to be fearless and undisturbed in sudden alarms than to be so in those that are foreseen; for it must have proceeded more from a state of character, because less from preparation; acts that are foreseen may be chosen by calculation and reason, but sudden actions must be in accordance with one's state of character.[50]

Men act bravely, not because they are compelled to but because it is noble for them to do so and they have developed this nobility. Women were of such low social status that such considerations did not apply to them.[51]

Inculcating norms of bravery at young ages thus could counteract natural dispositions toward fear. Norms associated with manliness, in particular, produced strong constraints over demonstrations of overt fear. The courageous man did not flee from dangerous situations as a coward or effeminate person would, because he was imbued with noble ideals about how to respond to danger. Courageous behavior, therefore, was more of an acquired character trait than a choice based on calculation and reason or an inborn trait.

The Classical age in Greece was followed by the Hellenistic period, which was marked by especially rapid changes that created social insecurities and undermined cultural assumptions. This period of transition and crises, from the late fourth to the second or first century BCE, led familiar worlds to collapse, the inherited fabric of beliefs to decay, and anxiety to diffuse broadly across the society. People sought reassurance in new philosophies that were specifically directed toward controlling anxiety.

Epicureans and Stoics, in particular, strove to achieve serenity through eliminating anxiety. In contrast to the social emphasis of the Classical philosophers, Epicurus (341–270 BCE) focused on teaching individuals how to maintain their peace of mind. His goal was to produce a state of tranquility, which he defined as "the state wherein the body is free from pain and the mind from anxiety."[52] The Epicureans promoted methods that achieved serenity through banishing passions, among which they counted anxiety, from human experience. They developed doctrines and techniques that allowed people to overcome their natural anxieties, especially anxiety stemming from fear of death or of the gods. Because they saw anxiety as arising from the discrepancy between desires and the capacity to reach them, they thought that the most effective way to control this emotion was to inhibit desires.[53] Epicureanism represented an important turn away from communal toward private and personal means of suppressing anxiety.

Two hundred years later, Epicurus's disciple Lucretius (99–55 BCE) also placed the control of fear at the center of his influential philosophy. The most important aspect of a happy life, he wrote, was to confront and conquer fears and unappeasable desires, in order to achieve a state of peacefulness (*ataraxia*). This is achieved by learning to distance oneself from worldly concerns:

> It is comforting, when winds are whipping up the waters of the vast sea, to watch from land the severe trials of another person: not that anyone's distress is a cause of agreeable pleasure; but it is comforting to see from what troubles you yourself are exempt. It is comforting also to witness mighty clashes of warriors embattled on the plains, when you have no share in the danger. But nothing is more blissful than to occupy the heights effectively fortified by the teaching of the wise, tranquil sanctuaries from which you can look down upon others and see them wandering everywhere in their random search for the way of life, compet-

ing for intellectual eminence, disputing about rank and striving night and day with prodigious effort to scale the summit of wealth and to secure power.[54]

Disengaging oneself from external sources of worry replaced demonstrating courage when facing danger as the ideal men should strive to reach. A state of mind that earlier periods might have characterized as cowardly became a valued goal.

For the Stoics fear was among the four major classes of emotions, the others being distress, desire, and pleasure. The Stoics' major goal was to teach people how to decrease their natural anxieties by reducing investments in achievements and goals. Mastering passions, accepting fate, and renouncing desires that could not be satisfied provided the best pathway to happiness. For example, Epictetus provided a typically Stoic solution for social anxiety: "I am in his power who can gratify my wishes and inflict my fears. Not to be a slave, then, I must have neither desire nor aversion for anything in the power of others."[55] He also gave perhaps the clearest depiction of the precedence of the unthinking power of fear, as well as the subsequent ability of reason to control fear:

> Mental "impressions," through which a person's mind is struck by the initial aspect of some circumstance impinging on the mind, are not voluntary or a matter of choice, but force themselves upon one's awareness by a kind of power of their own. But the "assents" through which those same impressions are cognized are voluntary and happen by one's own choice. That is why, when some terrifying sound occurs, either from the sky or from the collapse of a building or as the sudden herald of some danger, even the wise person's mind necessarily responds and is contracted and grows pale for a little while, not because he opines that something evil is at hand, but by certain rapid and unplanned movements antecedent to the office of intellect and reason. Shortly, however, the wise person in the situation "withholds assent" from those terrifying mental impressions; he spurns and rejects them and does not think there is anything in them which he should fear.[56]

Alarm in the face of impending danger was a natural emotion that was very difficult, although not impossible, to overcome. The Stoics tried to instill cognitive regimes that made people mentally rehearse confrontations with the things they feared most intensely and therefore create a readiness to deal effectively with them when they appeared.

Socialization from a young age and religious and philosophical belief systems were the most common ways of preventing anxiety in these ancient societies. Specifically medical responses to anxiety were rare. In general, Hippocratic physicians scorned heroic medical interventions into natural processes and acted in accordance with the famous dictum "First, do no harm." They believed that states of disequilibrium were correctable through changes in diet, exercise, sexual activity, and sleep. They also employed somatic treatments, including opium, which could cool an overheated brain and produce a calm, tranquil state. Alcohol, as well, was widely prescribed by physicians to alleviate anxiety. Overall, Hippocratic physicians strove to maintain or restore health by urging restrained modes of living.[57]

The Corybantic Rites, which Plato described as dance and musical treatments aimed at curing phobias and other anxious feelings, were an exception to the general principle of therapeutic nonspecificity. These rites, according to Classical scholar Ivan Linforth, were directly aimed at certain types of anxiety conditions: "The disorder that they cure is an obsession of emotions. It takes the form of mental agitation, characterized by terror and madness, and is accompanied by violent pounding of the heart."[58] They involved group dancing accompanied by music, especially flute music; they aroused the emotions of listeners, created frenzied states, and often resulted in a loss of consciousness of everything but the ecstatic rhythms of the dancing and music. When they ended, participants were supposed to emerge feeling calm and tranquil, cured of their fears, apprehensions, pounding of the heart, and other anxious symptoms. Such musical treatments persisted into the Roman period.[59]

CONCLUSION

Ancient conceptions of anxiety combined intuitions about its universal nature with culturally specific value systems. They recognized that anxiety was a natural emotion but they usually placed it in a context of moral virtues: men were honorable or dishonorable according to their ability to control their anxiousness. Anxious states were evaluated through their conformity to cultural norms that valued courage and scorned cowardice. While Classical medicine and philosophy emphasized that anxiety inevitably accompanied threatening and dangerous situations, they also recognized disordered anxiety, which was directed at culturally inappropriate sources of fear or arose without any apparent reason. Anxiety disorders were rarely individualized but were usually relatively nonspecific aspects of broader melancholic conditions accompanied by a variety of other psychological and physiological symptoms. Classical methods for controlling anxiety stressed regimens that trained youth to develop character traits that enabled them to overcome their natural fears. They also encompassed general changes in lifestyle, adherence to belief systems devoted to tension reduction, and, more rarely, medical treatments.

Subsequent to the Greeks and Romans, virtually no new developments occurred in medical thinking about anxiety until the end of the eighteenth century. Medical commentators throughout the intervening period relied on Greek physicians, especially Galen, as authorities on mental illness.[60] Galen's humoral theory, itself based on Hippocratic tenets, remained the basis for explanations of melancholia and other diseases for 1500 years. Likewise, the association of fear and sadness under the general melancholic umbrella persisted for centuries. Moreover, Hippocratic preferences for altering lifestyles continued to prevail over more intensive medical interventions. Correspondingly, Aristotelian conceptions of fear dominated philosophical discourse until the sixteenth century. Classical notions of fear and anxiety have had the most enduring impact of any conceptions yet developed.

CHAPTER THREE

From Medicine to Religion— and Back

FROM MEDICINE TO FAITH

The Roman emperor Constantine's conversion to Christianity in the early fourth century heralded a new era in the West, marked by a religious worldview that largely supplanted empirical conceptions of anxiety and other mental illnesses. The Christian emphasis on the immateriality and immortality of the soul thoroughly contrasted with the empirical tradition of Hippocrates and Galen. It transformed organic and empirical conceptions of disease to views grounded in faith, sin, and divine will. The moral view of anxiety shifted from its reflecting cowardice or courage to its indicating belief or impiety. Ideas crediting supernatural sources for both mental afflictions and salvation from them largely replaced not just medical but also the Epicurean and Stoic approaches, which advocated controlling anxiousness through limiting desires.

During the era between the fall of the Roman Empire and the Renaissance, people certainly had much to be anxious about: imminent or present epidemics of disease, famine, and other disasters created enormous insecurity. People interpreted such catastrophes as manifestations of God's wrath, not as natural forces. Indeed, the material world itself masked a more important and less transitory reality of either everlasting salvation or everlasting damnation. During this period the sources of anxiety were redirected from the body, public opinion, and the material world to what awaited people in the hereafter.[1]

St. Augustine (354–430), the most influential theological writer in this era, emphasized that relief from anxiety stemmed from faith in the teachings of Jesus Christ, a result he himself had experienced: "I neither wished nor needed to read further. At once, with the last words of this sentence, it was as if a light of relief from all anxiety flooded into my heart. All the shadows of doubt were dispelled." Augustine's notion that "our heart is restless, until it repose in Thee" epitomized the idea that faith in God was the best therapy for anxiety.[2] However, religion was not only the remedy for, but also the cause of, anxiety. Belief in God and an eternal afterlife could relieve anxiety but at the same time lead to tremendous uncertainty. Preoccupations with whether one was a member of the elect chosen to enter the kingdom of heaven and guilt over the consequences of sinning were potent sources of anxiety. Fear of perpetual damnation in the afterlife was a particular source of terror that persisted through the Reformation in the sixteenth century. Medicine had neither the cultural authority nor the practical skills to override the dominant spiritual interpretations of anxiety-producing situations during this period.

The millennium from about the fifth century CE to the sixteenth century saw not only a change from an empirical to a spiritual perspective on disease but also a more general denigration of the status of medicine. And, just as religious and magical concepts of healing replaced the naturalistic theories of Hippocrates and Galen, the practice of medicine itself was subjected to ecclesiastic control.[3] The church emphasized the salvation of souls more than the healing of bodies, so priests took primacy over doctors. Based on accounts of faith-based cures in the Gospels, patients required religious piety and healers needed divine powers. Witches and devils also had potent influences on illnesses in an era when mystical views predominated.

THE PERSISTENCE OF THE HIPPOCRATIC CORPUS

Although physicians became subordinate to theologians and medical thought itself languished for over a thousand years, the basic beliefs of what remained of elite medicine did not change. Indeed, from Galen's time through the Renaissance, the psychiatric corpus remained virtually undisturbed. The Hippocratic and Galenic tradition generally persisted in medical texts: anxiety continued to be yoked to depression and submerged in the general category of melancholia. For example, during the eleventh century, the Persian physician Avicenna (980–1037), author of the influential *Canon of Medicine*, defined the signs of melancholy as "bad judgment, fear without cause, quick anger, delight in solitude, shaking, vertigo, inner clamor, tingling, especially in the abdomen."[4] Later, in his influential *Three Books of Life*, Marsillo Ficino (1433–1499) also linked anxiety with depression, characterizing the typical state of melancholics as: "We hope for nothing, we fear everything."[5]

Renaissance physicians retained Hippocratic conceptions of melancholy. Felix Platter (1536–1614), for instance, highlighted the role of anxiety within melancholic conditions, calling melancholy a "kind of mental alienation, in which imagination and judgement are so perverted that without any cause the victims become very sad and fearful."[6] Fear was still a chief characteristic of melancholic conditions, which were mental disorders when they were "without cause."

The humoral theory of disease also endured in medical understandings and treatments of illness until the end of the seventeenth century and sometimes beyond. Humoral thought was foundational not only in the culture of physicians but also in the medical lore of common people and the lay healers who often treated them. The concepts persisted that each humor was associated with a certain temperament, each temperament was associated with specific diseases, disease resulted from imbalances among the various humors, and treatments aimed to correct such imbalances and restore the body to appropriate equilibrium. The prominent treatments for anxiety remained fresh air, exercise, good habits of sleeping, eating, and elimination, and control of the passions. Such treatments were typically intertwined with religious, magical, and folkloric remedies.[7]

THE ANATOMY OF MELANCHOLY

At the beginning of the seventeenth century, melancholia was the most prominent category of mental disturbance. It continued to be a capacious condition that encompassed a mixture of anxious and depressive symptoms. English vicar Robert Burton's encyclopedic *The Anatomy of Melancholy*, initially published in 1621, culminated a two thousand-year-old tradition that began with the Hippocratic corpus. It was not only the major compendium of literature about this condition but also the most comprehensive description of melancholy ever amassed. Running to nearly 1500 pages, Burton's magisterial work surveyed the entire sweep of writing on melancholy, beginning with the Bible and ancient Greeks and Romans and ending with the contemporary seventeenth-century studies. Perhaps because melancholy later became identified with a type of depressive condition, Burton's work is generally situated in the history of depression,[8] yet he was also centrally concerned with anxiety.

For Burton, the anxious component of melancholy was a fundamental aspect of human nature. "Great travail is created for all men, and an heavy yoke on the sons of Adam, from the day that they go out of their mother's womb, unto that day they return to the mother of all things. Namely, their thoughts and fear of their hearts, and their imagination of things they wait for, and the day of death." He regarded anxiety as a universal affliction that touched everyone: "From him that sitteth in the glorious throne, to him that sitteth beneath in the earth and ashes; from him that is clothed in blue silk and weareth a crown, to him that is clothed in simple linen."[9]

While melancholy was an integral aspect of normal life, it could sometimes indicate a form of mental disorder. Burton distinguished melancholic dispositions that were natural responses to troubled situations from melancholic habits that were mental disorders:

> Melancholy . . . is either in disposition or habit. In disposition, it is that transitory melancholy which goes and comes upon every small occasion of sorrow, need, sickness, trouble, fear, grief, passion, or perturbation of the mind, any manner of care, discontent, or thought, which causeth anguish, dullness, heaviness, and vexation of spirit. . . . And from these melancholy dispositions, no man living is free, no Stoic, none so wise, none so happy, none so patient, so generous, so godly, so divine, that can vindicate himself; so well composed, but more or less, some time or other, he feels the smart of it. Melancholy, in this sense is the character of mortality.[10]

In contrast, melancholic habits were "a kind of dotage without a fever, having for his ordinary companions fear and sadness, without any apparent occasion." Burton emphasized: "'Fear and sorrow' make it differ from madness; 'without a cause' is lastly inserted, to specify it from all other ordinary passions of 'fear and sorrow.'"[11] "Without any apparent occasion" and "without a cause" indicated the distinction Burton made between appropriate melancholic dispositions and disordered forms of melancholy.

Fear and sorrow were the two most characteristic symptoms of melancholy; Burton typically combined them in his descriptions of the condition. "Cousin-german to sorrow is fear," he noted, "or rather a sister, *fidus Achates* [trusty squire], and continual companion, an assistant and a principal agent in procuring of this mischief; a cause and symptom as the other." Clusters of symptoms marked by fear, anxiety, and apprehension were often indistinguishable from moods of sadness, despondency, and despair. Yet, Burton also noted that fear and sorrow could occur independently. Quoting Mercatus, Burton observed: "Fear and sorrow are no common symptoms to all melancholy; 'Upon more serious consideration, I find some' (saith he) 'that are not so at all. Some indeed are sad, and not fearful; some fearful, and not sad; some neither fearful nor sad; some both.'" Burton stressed that anxiety could present itself in many manifestations, including head and stomach aches, heart palpitations, trembling, sweating, and blushing.[12]

Almost any sort of future evil could cause fearful forms of melancholy. Indeed, Burton noted that there was an "infinite" variety of fears. He catalogued many of these specific fears:

> Fear of devils, death, that they shall be so sick, of some such or such disease, ready to tremble at every object, they shall die themselves forthwith, or that some of their dear friends or near allies are certainly dead; imminent danger, loss, disgrace still torment others, etc. . . . . Montanus speaks of one "that durst not walk alone from home, for fear he should swoon or die." A second "fears every man he meets will rob him, quarrel with him or kill him." A third dares not venture to walk alone for fear he should meet devil, a thief, be sick.

Of all these fears, the apprehension of some terrible object that presented an imminent danger was the "most pernicious and violent." Indeed, some of these objects, he recorded, could be so frightening to a person that they created terrors from which the victim never recovered. Such cases represented melancholic disorders "with cause."[13]

Anxiety had an especially intimate link to love: "Every poet is full of such catalogues of love-symptoms; but fear and sorrow may justly challenge the chief place . . . love melancholy . . . 'Tis full of fear, anxiety, doubt, care, peevishness, suspicion; it turns a man into a woman . . . ." Burton went on to note: "Now if this passion of love can produce such effects if it be pleasantly intended, what bitter torments shall it breed when it is with fear and continual sorrow, suspicion, care, agony, as commonly it is, still accompanied! What an intolerable pain must it be!"[14]

Using cases from the Romans as examples, Burton also discussed social phobias: "It amazeth many men that are to speak or show themselves in public assemblies, or before some great personages; as Tully confessed of himself, that he trembled still at the beginning of his speech; and Demosthenes, that great orator of

Greece, before Philippus." He also characterized a case resembling agoraphobia: "If he be in a throng, middle of a church, multitude, where he may not well get out, though he sit at ease, he is so misaffected." Such a person "dare not come in company for fear he should be misused, disgraced, overshoot himself in gesture or speeches, or be sick; he thinks every man observes him, aims at him, derides him, owes him malice."[15]

Burton also described panic attacks: "Let them bear witness that have heard those tragical alarums, outcries, hideous noises, which are many times suddenly heard in the dead of night by irruption of enemies and accidental fires, etc. those panic fears, which often drive men out of their wits, bereave them of sense, understanding, and all, some for a time, some for their whole lives, they never recover it."[16] Indeed, Burton identified most of the anxiety conditions, including specific and social phobias, agoraphobia, panic, and generalized anxiety disorder, which remain in the contemporary psychiatric canon regarding anxiety. His Hippocratic emphasis on fear without a cause also provided a template with which to separate anxiety disorders from contextually grounded normal fears.

Burton's recommendations for treatment generally echoed the tenets of Hippocratic medicine: restoring balance to diet, exercise, surroundings, sleep, and emotion. He also indicated a number of more specific remedies for fearful types of melancholy, chief among which was alcohol. "A cup of wine or strong drink," he observed, "takes away fear and sorrow."[17]

Burton's rambling and disjointed style, not to mention the inconsistencies and lack of systemization that plague his work, meant that his vast compendium was not a useful foundation for the more scientific study of anxiety that would emerge in later centuries. It remained for future diagnosticians to build on Burton's work, to disentangle anxious from depressed conditions, and to make anxiety conditions in themselves a focus of medical concern.

RICHARD NAPIER

A remarkable collection of case histories of anxious conditions is found in the records of 767 people who sought help from one seventeenth-century English physician and clergyman, Richard Napier, the subject of Michael MacDonald's book, *Mystical Bedlam*. These accounts vividly show the prominent role of normal and, more rarely, disordered anxiety in the complaints that sufferers brought to this general practitioner.[18]

Common people during this period had much to worry about. Debt and the consequent fear of poverty was the greatest single source of anxiety. Men, in particular, worried about the lack or loss of money. Fear of disease was also a prominent concern: "Seventeenth-century Englishmen were death's familiars, for epidemics, consumption, parasites and dysentery, accidents, infections, and botched childbirths killed children and adults, family and friends, earlier and more suddenly than the diseases we dread today." Fears of bewitchment or witchcraft accusations were also acute sources of anxiety within the closed, intimate world in which these villagers resided. The perceived witch was likely be an everyday presence in one's life and impossible to avoid. Such fears bedeviled about two-thirds of Napier's patients. Anxiety over sins that could lead to eternal damnation and religious despair also plagued these sufferers.[19]

Problems involving courtship and married life, especially lover's quarrels, unrequited love, parental objections to a love object, and double dealing were another central concern. Such fears afflicted nearly 40 percent of Napier's clients. Among patients complaining of marital problems, 84 percent were women (then, as now, about two-thirds of all patients were women). Age was also pertinent; young adults between 20 and 29 composed a disproportionate number of Napier's patients. Uncertainties over their futures were prominent reasons for seeking advice. "Many of these young people complained to their physician about the anxieties of courtship and marriage and the uncertainties of getting a living and bearing children, problems that accompanied the transition

from youthful dependence on parents and masters to full independence as married adults."[20]

The anxiety of Napier's patients was usually linked to the normal uncertainties of existence, not to madness or melancholia.
Worries about courtship, marriage, death, disease, impoverish-
ment, witchcraft, and damnation were realistic sources of concern
among villagers facing both present threats and uncertain futures.
Therapies for these conditions were nonspecific: "Napier's anxious
patients received similar treatments as those with other problems.
The regimen of treatment for mad and troubled patients differed
little from the measures used to cure other patients. Regardless
of their symptoms, almost every one of Napier's mentally disturbed patients was purged with emetics and laxatives and bled
with leeches or by cupping."[21] Typical of medical practitioners at
the time, Napier combined religious and magical with medical
approaches, using prayer and protective amulets to help heal his
patients.

Most anxious conditions he considered worthy of treatment
were not mental disorders. The complaints of some of Napier's
anxious patients, however, went beyond ordinary anxiety and
seemed to be disordered, because they were not proportionate to
the patient's actual circumstances. MacDonald describes some of
these cases:

> Everybody was afraid occasionally of the perils that many of Na-
> pier's melancholy patients complained about: devils, death, ill-
> ness, accident, disgrace, robbery, and witches, for example. The
> apprehensions of patients . . . that they would die or be driven
> mad by disease were tokens of melancholy because they did not
> seem ill enough to justify such fears. . . . Ann Wilson thought
> that the medicine a physician had given her harmed her un-
> born child. The idea that women should avoid dosing themselves
> with physic during pregnancy was commonplace; Wilson's fear-
> fulness was melancholy because it persisted even after the child's
> healthy birth had vindicated the careless physician. . . . Similarly,
> although violent crime was endemic, persons . . . who claimed

that some unknown malefactor was going to kill them, turned a legitimate apprehension into a melancholy fear by detaching it from any plausible situation or dangerous enemy.[22]

These cases indicated melancholic disorders because they could not be encompassed within the commonsense assumptions of village culture. Such deranged anxiety conditions seem to have been relatively rare among this group. While it is impossible to know how similar Napier's patients were to others during this period, it seems very likely that they were representative and so reflected the central role of normal anxiety among patients of general practitioners more broadly.

ANXIETY IN PHILOSOPHY

Fearful emotions were of central concern in philosophy as well as in medicine. In the sixteenth century French essayist Michel de Montaigne (1533–1592) famously noted: "The thing I fear most is fear. Moreover, it exceeds all other disorders in intensity." He pointed out that fear was such a powerful emotion it could lead to madness: "In truth, I have known many people to become insane from fear." Montaigne emphasized how the instinctual nature of fears overpowered cognitive controls: "Put a philosopher in a cage of thin iron wire in large meshes and hang it from the top of the towers of Notre Dame of Paris; he will see by evident reason that it is impossible for him to fall, and yet . . . he cannot keep the sight of this extreme height from terrifying and paralyzing him."[23] Montaigne noted that, although he himself was only "moderately frightened of heights," he shivered and trembled when on top of a mountain, even when he was well away from the edge of the cliff and couldn't possibly fall off of it.

Montaigne might have been surprised at the number of Napier's villagers who suffered anxious fears. He associated fear with the higher orders of society, remarking that fear was more likely to characterize the wealthy than those who had nothing to lose: "Those who are in pressing fear of losing their property, of being exiled, of being subjugated, live in constant anguish, losing even the capacity to drink, eat, and rest; whereas the poor, the exiles, and the slaves often live as joyfully as other men."[24]

The political philosopher Thomas Hobbes (1588–1679) likewise admitted, "The strongest emotion I ever felt was fear."[25] Hobbes's innovation was to focus on the social nature of fear instead of on fearful individuals. He believed that society was bound together through "mutual fear": men not only feared others but also realized that others feared them. Society itself originated in fear, because the fear each member had of others led men to create government to protect them from the acts of others.

The radical thought of René Descartes (1596–1650) was perhaps the most influential philosophical stimulus to subsequent thinking about anxiety. Descartes was instrumental in moving philosophy from a theological to an empirical grounding, emphasizing the incommensurability of mind and matter. In contrast to accepted theological beliefs, but in concordance with the emergence of Newtonian science, he viewed the body as a mechanism separate from the incorporeal nature of the mind. Like all matter, bodies acted according to quantifiable laws of nature.

Cartesian conceptions had fundamental implications for the study of mental disorders because minds, unlike matter, could not be diseased; genuine mental illnesses, therefore, must be located in the body and, more specifically, in the brain. This placed such diseases in the medical, not the theological, realm. Despite Descartes's own belief that his split of mind and matter protected the sanctity of faith from scientific findings, his unchaining of material from spiritual entities challenged deeply rooted religious beliefs and seemed dangerously atheistic to the theological authorities of his time.

In contrast to Descartes, other philosophers rooted emotions such as fear and anxiety in the mind rather than the body. In particular, John Locke (1632–1704) emphasized how mentally ill people joined false ideas together and so made incorrect deductions. Learned associations of ideas could lead to irrational connections between perceptions of conditions, such as darkness, and resulting emotions, such as the fear of goblins. Likewise, Locke's view that madness was more a defect in understanding than of either the body or the soul led him to emphasize that education could lead the deluded to think correctly and thus be cured of their fears:

> If your child shrieks and runs away at the sight of a frog, let another catch it and lay it down at a good distance from him; at first accustom him to look upon it; when he can do that, to come nearer to it and see it leap without emotion; then to touch it lightly, when it is held fast in another's hand; and so on, until he can come to handle it . . . .[26]

This advice, as one current writer about phobias notes, "could have come straight out of a modern behavior therapy textbook."[27] Locke's theories placed the study of mental illness squarely into the arena of disordered mental, not organic, processes.

Scottish philosopher David Hume's (1711–1776) *A Treatise on Human Nature* (1739) presented a fresh analysis of fear and anxiety.[28] Before Hume, anxiety and sorrow had been interconnected for thousands of years as the core symptoms of melancholia. Hume gave a penetrating analysis of how isolatable characteristics of anxiety distinguished it from other emotions, including sadness. He considered fear and hope, along with grief and joy, as the basic emotions (what Hume called "direct passions"). The certain occurrence of good things produced joy while definite evils led to grief or sorrow. Uncertainty that was associated with positive events led to hope. Fear arose when evil events were uncertain. The core difference, then, between grief and fear was that grief resulted from the actual occurrence of distressing things while fear was connected to uncertainty and anticipation over whether some evil event might happen.

For Hume, fear, and its correlate states of anxiety, consternation, and terror, was a core aspect of human experience. Malevolent events that were probable, but not assured, led to fear: "Tis evident that the very same event, which by its certainty wou'd produce grief or joy, gives always rise to fear or hope, when only probable and uncertain." Hume provided the example of a father who hears that one of his sons has been killed but does not know which one. If the man knew with certainty which of his sons had died, he would feel grief, not fear. The uncertainty leads him to experience fear rather than grief. All sorts of events that were uncertain, even pleasurable ones, could be sources of anxiety. Hume used the example of a virgin on her wedding night who anticipates intense joy but nevertheless goes to bed feeling fearful and apprehensive. Because uncertainty was such a pervasive aspect of experience, reasoned Hume, fear was perhaps the most basic human emotion.[29]

Hume was not concerned with the implications of his philosophical analysis for notions of melancholy. Moreover, he was interested in normal manifestations of fear, not in conceptions of anxiety disorders. Therefore, his ideas developed well outside of any medical context. Hume's concepts, however, provided a philosophical groundwork for the much later splitting of anxiety from other conditions that eventually penetrated psychiatric thinking.

ANXIETY IN SEVENTEENTH- AND EIGHTEENTH-CENTURY MEDICINE

A radical change in the Western intellectual tradition occurred in the seventeenth century as the inductive, empirical, and observational methods of Bacon and Newton overturned theological conceptions. By midcentury, medical views of mental illnesses had largely displaced faith-based ones among the educated public. Within medicine, after thousands of years of dominance by Hippocratic views of humoral imbalance, a new system arose that was based on disturbances of the brain and the nervous system. Notions of disease specificity also began to emerge, particularly in the work of English physician Thomas Sydenham (1624–1689). Sydenham proposed that each disease had natural forms with uniform presentations in different individuals, an idea distinct from holistic Hippocratic thought.

Symptoms of anxiety began to emerge from the capacious category of melancholia and to gain a place within the new organic category of "nervous disorders." The emergent theories of disease focused on the nervous system as the source of health and illness, emphasizing the importance of nerves, fibers, and organs. Accordingly, the causes of nervous conditions were found in physiology and, in particular, brain lesions. In this scheme, nervous disorders fell under the domain of general physicians, neurologists, and spa doctors, not of psychiatrists, who were expected to treat conditions of insanity.[30]

An English physician, Thomas Willis (1621–1675), who coined the term "neurology," strove to localize various mental functions to particular regions of the brain. He developed a theory of neurological activity based on the notion of "animal spirits"—life-carrying fluids that passed through the nerves and transmitted information between the sense organs, brain, and muscles. Although Willis used a traditional definition of melancholy that joined fear and sadness, he also partially differentiated the physiological roots of these two components: "First, in Sadness, the flamy or vital part of the Soul is straitned, as to its compass; and driven into a more narrow compass; then consequently, the animal or lucid part contracts its sphere, and is less vigorous; but in Fear both are suddenly repressed and compelled as it were to shake, and contain themselves within a very small space."[31] Willis enumerated the physiological manifestations of fear, including hair standing on end, a loosening of the nerves, involuntary excretion, and fainting. The concept of localized brain functions was instrumental in initiating the overthrow of the humoral theory and establishing a new organic paradigm. Soon, the Welsh physician Nicholas Robinson (1697–1775) was able to assert that "Every change of the Mind indicates a Change in the Bodily Organs."[32]

George Cheyne (1671–1743) was a society "nerve doctor" whose book *The English Malady* (1733) was probably the most influential popularization of the new conceptions of nervous disorders.[33] Cheyne's mechanistic model of the human organism focused on the nervous system, fibers, tissues, and physical organs of the body. The nerves mediated between the mind and the brain, possessing attributes of both.[34] They conveyed sensations throughout the body and carried messages among the brain, internal organs, and limbs and so were deeply intertwined with the stomach and brain. He subsumed a variety of conditions, including hysteria, hypochondria, and some cases of melancholy, under the category of nervous distempers. Cheyne rooted nervous conditions in the body rather than the mind, emotions, or soul, thus placing them within the purview of the physician rather than the clergyman or philosopher.

Cheyne's major impact in medical history was to increase the importance of nervous afflictions in the mind of a wider, lay audience. Nervous ailments, he pointed out, by no means represented minor complaints. Indeed, Cheyne stated, "of all the Miseries that afflict Human Life, and relate principally to the Body, in this Valley of Tears, I think *Nervous* Disorders, in their extream and last Degrees, are the most deplorable, and beyond comparison the worst."[35] Nevertheless, he explicitly distinguished nervous disorders from insanity and thus from serious melancholia, which continued to be seen as a type of madness.

Anxiety and dejection, along with a cornucopia of physical complaints, fell into Cheyne's expansive notion of what constituted a nervous disorder. He noted twenty symptoms of "nervous disorders," including "sinking, suffocating, and strangling," "fidgeting," and "peevishness." Cheyne also described how in his twenties he himself had been suddenly stricken by a condition of "Fright, Anxiety, Dread, and Terror" that led him to be bedridden for months.[36]

A unique aspect of Cheyne's work was that he rooted nervous conditions not just in individual bodies but also in the social, cultural, and historical forces that underlay a person's way of life. He thought that social conditions, especially prosperity, rapid social change, and dissolute lifestyles, had led to the epidemic of nervous afflictions among the English. Indeed, he believed that huge numbers of well-bred Englishmen suffered from "the English Malady": "These nervous Disorders being computed to make almost one third of the Complaints of the People of Condition in England."[37] This could be the first rudimentary statement of psychiatric epidemiology.

Cheyne attributed the English malady to the progress of civilization and especially to the lifestyles of the most prosperous residents of the country. Indeed, nervousness became a sign of social superiority. Conversely, those lower in the social hierarchy he considered largely immune to nervous disorders: "Fools, weak or stupid Persons, heavy and dull Souls, are seldom much troubled with Vapours or Lowness of Spirits." While Cheyne did not associate the characteristics of nervous disorders with men or women, he did link them to effeminate individuals of either sex.[38]

Cheyne's book was widely read, and it helped promote the notion of nervousness as a fashionable feature of well-bred gentlemen and ladies. By associating anxiety with social success, prosperity, and excessive consumption while at the same time providing it with a physiological grounding, he made nervous conditions socially acceptable.[39] Cheyne's view also moved nervous disorders away from the lonely, marginal, outsider status of melancholia and toward a sociable, integrated, and respectable insider status. It paved the way for subsequent notions of nervous diseases, including Beard's "neurasthenia" and Freud's "neuroses," which became not just tolerable but even fashionable illnesses among the bourgeoisie.

While Cheyne's book gained immense popularity among the educated public, the works of William Cullen (1710–1790), a Scottish professor of medicine and the founder of psychiatric theory and practice in Great Britain, had more influence on medical thought. Cullen's impact persisted through the many physicians he trained, and his ideas traveled to the United States through one of his students, Benjamin Rush (1745–1813), the founder of American psychiatry. Cullen's writings grounded mental disturbances in neurophysiology, specifically in states of increased and decreased activity of nervous power in the brain, which he called "excitement" and "collapse" respectively. Yet, his theory was not mechanistic and, following Locke, he emphasized how nervous activity could lead to inappropriate associations of ideas. Cullen's works thus served as a bridge between psychological and somatic views of mental illness.[40]

Cullen created the term "neurosis" to refer to various ailments of the central nervous system. In his conception, neurosis (or nervous disorder) was one of four classes of diseases. It, in turn, was subdivided into four subclasses, one of which was vesania (disorders of intellectual functions). Vesania itself had four categories, including melancholia, along with mania, amentia (mental retardation), and onierodynia (intense distress associated with dreaming).[41] Cullen might be the first "splitter" of psychiatric disorders, placing hysteria in the subclass of spasmodic affections (irregular motions of the muscles) and hypochondriasis in the adynamiae subclass (deprivations of involuntary motions) and distinguishing both disorders from melancholia. He differentiated melancholia from other types of disorders because it was "always attended with some seemingly groundless, but very anxious fears."[42] He also noted the presence of melancholic temperaments that were marked by cautious and fearful dispositions.

Cullen did not view the physical and psychic symptoms of anxiety as constituting a distinct disorder but as aspects of broader syndromes. He sometimes linked anxious symptoms with the ailments and worries of the hypochondriac, thus partially removing anxious symptoms from their association with melancholic fear and sorrow, which he considered a form of insanity. By associating some anxiety conditions with hypochondriacal states, which were not insanity, Cullen's nosology, like Cheyne's, helped pave the way for the later understanding of neurasthenic and neurotic disturbances.[43]

Throughout the eighteenth century, medical writing about mental illnesses focused on nervous disorders. Scottish physician Robert Whytt (1714–1766) emphasized the protean nature of this subject, noting, "Those morbid symptoms which have been commonly called *nervous*, are so many, and so various, and so irregular, that it would be extremely hard, either rightly to describe or fully to enumerate them." Many of the symptoms of these diseases, Whytt noted, were somatic, including "wind in the stomach and intestines, heart-burning, an uneasy, though not painful sensation about the stomach," and "palpitations, or trembling of the heart," while others were psychological, such as "low spirits, anxiety, and sometimes great timidity."[44]

William Battie's *Treatise on Madness*, published in 1758, provided the first sustained treatment of anxiety from a psychiatric viewpoint. Battie, the superintendent of St. Luke's Hospital in London, might be the first physician who defined anxiety as a distinct mental disorder. He distinguished anxiety from madness, noting, "Anxiety is no more essentially annexed to Madness . . . than Fever, Head-ach, Gout, or Leprosy." Moreover, Battie indicated that anxiety could be adaptive, pointing out that it was "absolutely necessary to our preservation, in such a manner, that without its severe but useful admonitions the several species of animals would speedily be destroyed."[45]

Some forms of anxiety were "original," wrote Battie, and stemmed from the inherited constitution of sufferers who had oversensitive nerves. This type of anxiety, arising from a weakness of the nerves that resulted from obstruction of the blood vessels, was unlikely to be curable. In contrast, "consequential" forms of anxiety resulted from some "intolerable impulse of external objects" that created "too great or too long continued force."[46] While there was little hope of remedying original forms of anxiety, those who suffered from consequential anxiety could benefit from eating simple food, breathing clean air, engaging in nonstressful employment, and adopting moderate lifestyles.

Throughout the eighteenth century, psychiatrists specialized in dealing with psychotic states associated with madness and did not treat nervous disorders. Instead, general physicians, as well as a plethora of medically marginal herbalists, faith-healers, charlatans, and religious healers, provided care to people with anxious conditions.[47] Indeed, the label "nervous disorder" was a preferred diagnosis because it was *not* connected with a profession that dealt with madness. Despite the displacement of Hippocratic explanations of mental disturbance, the Hippocratic emphasis on moderation in all aspects of lifestyle continued. As in medicine more generally, opium—which was readily available and widely used—was also a common treatment for anxious conditions.[48]

Many observers, following Cheyne, believed that the eighteenth century was associated with an epidemic of nervous disease. Indeed, historian Edward Shorter notes a competition in the Dutch city of Utrecht for the best essay on the topic "the causes of the increasing nervous disease in our land."[49] One of Cullen's students, Thomas Trotter (1760–1832), observed, "At the beginning of the nineteenth century, we do not hesitate to affirm, that *nervous disorders* have now taken the place of fevers, and may be justly reckoned two thirds of the whole, with which civilized society is afflicted."[50] Nervous conditions seemed to be ubiquitous in Western countries during this period, at least among the well-bred.

CONCLUSION

Between the fourth and sixteenth centuries, religious, magical, and folkloric views of anxiety and other mental conditions largely displaced empirically based Hippocratic and Galenic conceptions in Western societies. While the latter beliefs persisted within medicine through the medieval period, medical knowledge itself was overridden by ecclesiastical strictures. Over the course of the seventeenth century, however, anxiety was once again increasingly likely to be viewed within a medical, as opposed to spiritual, framework. Within medicine, the influence of humoral pathology, which had dominated medical thinking since Hippocratic times, gradually waned. By the end of the eighteenth century, while humoral conceptions remained popular theories of temperaments in the general culture, medicine preferred physiological accounts to explain how mental disturbances resulted from malfunctioning nervous systems.

Alongside somatic explanations that stressed the activities of the nervous system, the Lockean notion persisted that psychiatric conditions resulted from faulty associations of ideas. Many factors, including individual temperament, heredity, and external circumstances, were considered to be related to the faulty working of nerves.[51] Psychological explanations and treatments thus vied with physiological ones, especially as moral therapies, which emphasized the social roots of mental disturbances, became dominant within asylums.[52]

Despite the growing interest in nervous conditions, classifications of mental disorders remained rudimentary. At the end of the eighteenth century, Cullen notwithstanding, most categorizations remained very general and continued to submerge anxiety within the amorphous category "nervous disorders." However, the emergence of the concept that the nervous system was the locus of mental conditions paved the way for a widespread medicalization and specification of anxiety disorders during the nineteenth century. Physicians and psychiatrists began to differentiate the psychological and the somatic components of anxiety and to disentangle anxiety conditions from other conditions. By the end of the nineteenth century, anxiety was regarded as a distinct, and important, condition within the psychiatric corpus.

CHAPTER FOUR

The Nineteenth Century's New Uncertainties

❖ ❖ ❖

Current Western conceptions of anxiety were shaped in the nineteenth century. During this period new technologies, which created railroads and factories, powered unprecedented rates of social change. These changes in turn created mass migration, urbanization, and industrialization that tore individuals from more settled ways of life. The newly industrialized world created a wide range of uncertainties and at the same time weakened traditional social and cultural supports. Contemporary political, philosophical, scientific, and artistic ideologies, which generally emphasized individual needs and rights, supplanted previously prevailing religious beliefs among large segments of Western societies.

A passion for measurement also marked the nineteenth century. Positivism, whose dictum was "to measure is to know," became the dominant philosophy of science. Scientific medicine made great progress, developing the capacity to detect such phenomena as the concentration of chemicals in blood and urine, body temperature, and blood pressure and to correlate them with demographic factors such as sex, age, and physique. Physicians were beginning to understand diseases at the level of tissues and cells, so they could define specific organic conditions much more sharply. By the middle of the century, neurologists were using microscopes to observe nerve fibers.[1]

A new, scientific medical vocabulary had spread rapidly during the eighteenth century, largely replacing humoral and religious terminology. By the beginning of the nineteenth century, William Cullen's term "neurosis" had replaced the eighteenth century's moniker "nervous disease." "Neurosis" was still a very general label; it embraced all disorders of the nervous system in which no physical lesion had yet been found. Around this time doctors began to dominate the treatment territory previously occupied by priests and theologians. French psychiatrist Phillipe Pinel (1745–1826), for example, described the case of a young monk who was struggling with his intense, but forbidden, sexual desires. Although the monk considered his condition to be a symptom of spiritual imperfection, Pinel provided a diagnosis of nervous illness.[2]

The speed of social change, rise of individualism, and belief in scientific medicine, coupled with the decline of traditional meaning systems, provided fertile soil for a growing medicalization of anxiousness. The rising bourgeoisie, especially the wives and daughters of the wealthy middle class, was ready for fresh ways of seeing anxious symptoms. These were people who did not have many worries about experiencing famine, premature death, or economic ruin and so could direct their anxiousness to social and psychological insecurities. As the nineteenth century progressed, laypeople became increasingly likely to self-diagnose and seek medical attention for nervous conditions. Claims that these ailments were genuine diseases had special appeal for the higher classes, who could be assured that their suffering was real and deserving of medical treatment yet also distinct from insanity. Unlike melancholia or mania, "nervous problems" were not viewed as signs of madness or as grounds for commitment to a mental institution.

Before the nineteenth century, psychiatry as a medical specialty barely existed outside of those physicians who practiced in and ran inpatient asylums; the earliest psychiatrists did not handle commonplace psychological complaints. Responsibility for treating such afflictions was widely diffused through society, encompassing physicians, spiritual healers, herbalists, apothecaries, alchemists, and astrologers. Especially during the latter half of the century, a new specialty, "nerve doctors," who focused on diagnosing and treating nervous complaints, grew and flourished. These practitioners disassociated themselves from mental institutions and psychotics and developed community-based practices that treated the generally well-off sufferers of nervous complaints.[3]

Physicians who treated anxious patients in private outpatient practices faced the dilemma of how they should classify the ailments they confronted. They required a medically based model to assure help seekers that their symptoms were organic and not imagined, yet the huge variety of conditions, coupled with the idiosyncratic nature of the individual complaints, posed a formidable obstacle to creating a scientific classification scheme. Terms such as "nervous collapse" or "nervous exhaustion" did not distinguish acute from chronic conditions, different levels of severity, or distinct causes. Nevertheless, over the course of the century demand soared for medical-sounding names for anxious symptoms that didn't require hospitalization.

CONNECTING ANXIETY TO THE BODY

Prevailing conceptions of anxiety at the beginning of the nineteenth century were nonspecific and enmeshed in broader categories of mental illness. Both the psychic and the somatic aspects of anxiety continued to be grouped within broader concepts that nervousness resulted from brain disturbances.[4] The grounding of anxious conditions in somatic states, which began during the eighteenth century, became firmly entrenched in nineteenth-century medicine. Physical lesions were the only plausible scientific explanation of mental afflictions at the time. Historian Janet Oppenheim notes that "Victorian alienists," who treated hospitalized mentally ill patients, "believed that physiological explanations were needed to give psychiatry the authority of science in an era when physics, chemistry, geology, biology, and astronomy appeared to be uncovering the secrets of the universe."[5] In particular, the discoveries of French anatomist Paul Broca (1824–1880) seemed to open opportunities to localize disorders in particular regions of the brain.

The most influential psychiatrists of the time, such as Wilhelm Griesinger (1817–1868), focused on organic conditions, which they believed were largely inherited. "Patients with so-called 'mental illnesses,'" Griesinger famously asserted, "are really individuals with illnesses of the nerves and brain."[6] While psychosocial causes could *produce* mental disorders, the resulting *conditions* were viewed as thoroughly somatic. As this theoretical approach spread, a much broader range of conditions began to be understood as reflecting somatic rather than mental processes. "In all cases where disorder of the mind is detectable," Scottish psychiatrist W. A. F. Brown (1805–1885) declared, "from the faintest peculiarity to the widest deviation from health, it must and can only be traced directly or indirectly to the brain."[7]

The growing use of drugs, stimulated by the discovery of the alkaloids at the beginning of the nineteenth century, reinforced somatic conceptions of mental illness. If consuming a medicine improved the condition, the reasoning went, then the malady must be somatic in origin. The consumption of bromides, morphine, quinine, codeine, and comparable drugs to treat psychological conditions soared over the course of the century.[8] Moreover, the opiates continued to be easily available for anyone who wanted them. All these drugs were used nonspecifically across the range of psychiatric conditions.

Charles Darwin's influential evolutionary theory also strengthened biological understandings of anxiety in medicine and psychiatry. Despite the fact that the genetic mechanisms responsible for the transmission of inherited traits would not be discovered until several decades later, Darwin emphasized the inherited basis of fear and other traits: "That the chief expressive actions, exhibited by man and by the lower animals, are now innate or inherited,— that is, have not been learnt by the individual,—is admitted by every one. So little has learning or imitation to do with several of them that they are from the earliest days and throughout life quite beyond our control."[9] Fear was not learned through experience but instinctually arose as a response to imminently threatening situations.

Darwin used his own experience to illustrate the involuntary aspect of fear:

> I put my face close to the thick glass-plate in front of a puff-adder in the Zoological Gardens, with the firm determination of not starting back if the snake struck at me; but as soon as the blow was struck, my resolution went for nothing, and I jumped a yard or two backwards with astonishing rapidity. My will and reason were powerless against the imagination of a danger which had never been experienced.[10]

Moreover, Darwin proposed the mechanism for why living creatures were naturally fearful: natural selection. Organisms that became anxious in the face of danger had a greater chance of surviving and reproducing than those that didn't.

Darwin showed not only that normal fear was an innate, inherited characteristic that arose involuntarily as a reaction to danger but also that fear had distinctive facial and postural characteristics in both humans and animals that distinguished it from other emotions, such as sadness, joy, and disgust: "Fear [in other animals] was expressed from an extremely remote period, in almost the same manner as it is by man; namely, by trembling, the erection of the hair, cold perspiration, pallor, widely opened eyes, the relaxation of most of the muscles, and by the whole body cowering downwards or held motionless."[11] Darwin did not discuss subjective experiences of fear but emphasized how its characteristic bodily signs served to convey a distinct emotional state to others.

Darwin's thought facilitated both the growing somatization and the heightened specificity that the study of anxiety underwent in the letter half of the nineteenth century. It also provided a theoretical explanation for the universality of fearful emotions and their usefulness in dealing with the dangerous contexts of life that Aristotle had described over two millennia before: creatures that developed fearful emotions in dangerous situations were better able to survive and reproduce than those that didn't. Consequently, they were more likely to pass to future generations whatever traits were responsible for the development of fear. Darwin's work remains a seminal influence on contemporary evolutionary psychiatry and psychology.[12]

In the middle of the century, nervous diseases began to split into two groups, "organic" and "functional." Organic conditions were linked to some specific bodily pathology; those for which no such basis could be found were placed in the functional group. By the 1840s, functional disorders were called "psychoneuroses"; they were nervous conditions that lacked any known grounding in anatomical lesions.[13] Most observers, however, believed that an underlying organic basis would eventually be found for functional states.

The somatic emphasis led to treatment of nervous disorders by a variety of nonpsychiatric medical practitioners, depending on the physiological manifestations of the condition. Pounding heart, breathing difficulties, and stomach pain were considered to reflect cardiovascular and gastrointestinal problems, respectively, and were treated by the appropriate specialist. For example, an American internist, Jacob Da Costa (1833–1900), described the condition "irritable heart" among soldiers, which referred to heart palpitations, chest pain, and extreme fatigue brought about by "nervous irritability."[14] Sufferers of this condition often believed that they were in danger of dying and suffered panic attacks as a result. Cardiologists, not nerve doctors, commonly dealt with anxious symptoms connected with heart or breathing problems. Similarly, an Austro-Hungarian neurologist, Moritz Benedikt (1835–1920), described one patient's state of dizziness whenever he entered a street or a square. Benedikt, in accordance with the dominant somaticizing tradition, attributed this condition to a disorder of the inner ear. Thus, inner ear specialists typically treated the dizziness associated with panic attacks.[15]

Even those anxious states that were viewed as within the realm of psychiatrists were generally regarded as epiphenomena of physiological states and not as psychological conditions in their own right. German psychiatrist Ewald Hecker (1843–1909), in writing one of the first descriptions of panic attacks, emphasized their somatic aspects: "Among neurasthenics, it happens with surprising frequency that instead of complete anxiety attacks a number of possible physical symptoms of anxiety may appear individually in pronounced attacks, without being accompanied by psychological feelings of anxiety."[16]

Although biological approaches dominated the study of anxious conditions throughout the century, a new paradigm was gradually beginning to emerge that focused on the psychic, as well as the somatic, aspects of anxiety. In 1872, German psychiatrist and neurologist Carl Westphal (1833–1890) developed a groundbreaking description of how the fear of entering public spaces caused dizziness. He coined the term "agoraphobia" (*die Agoraphobie*) to refer to an intense fear of going out in public that led sufferers to avoid such situations. Westphal differentiated this condition from psychoses, noting that "no hallucinations or delusions . . . cause this strange fear."[17] He emphasized how the source of agoraphobic fear was not streets and public spaces per se but rather the anticipation of the intense anxiety that entering them would provoke. His interpretation thus in part defined anxiety as a subjective state of consciousness instead of a physiological disturbance of the heart or brain. Nevertheless, Westphal was a thoroughgoing materialist and a fervent proponent of biological psychiatry, so, in line with the dominant somatic orientation, he also associated agoraphobia with the vertigo connected with the inner ear.[18]

In 1879, British psychiatrist Henry Maudsley used the term "panic" to describe states marked by sudden onset of extreme agitation, heart palpitations, trembling, and terror. In line with the view that mental disorders were brain diseases, Maudsley held, "Mental disorders are neither more nor less than nervous diseases in which mental symptoms predominate."[19] Despite the growing attention to the psychological aspects of anxious conditions, the description and understanding of them remained firmly grounded in physiology.

The focus on the somatic foundation of anxiety left unresolved the question of how to interpret subjective states of anxiousness. French asylum superintendent Benedict Morel (1809–1873) pioneered the view that both psychological and somatic symptoms of anxiety could stem from a single anatomical source. He described a condition of "emotional delusions" marked by subjective complaints of anxiety, phobias, and obsessions but also by cardiovascular, gastrointestinal, and nervous symptoms. Morel considered the ganglionar (autonomic nervous system) as the source of all these symptoms. His efforts represented one of the first attempts to classify disparate psychological and somatic anxiety conditions through a common etiological source.[20]

Under the sway of Cartesian philosophy, medical thinking emphasized that only bodies, not minds, could be diseased. Purely mental symptoms of anxiety were difficult to fit into this view, so medical practitioners tended to ignore psychological states that were not accompanied by physical manifestations. The treatment of souls belonged to religion. People had perennially sought consolation from clergy for anxiousness, and many people continued to believe that it had spiritual dimensions. Danish psychologist, philosopher, and theologian Sören Kierkegaard (1813–1855) extensively described the purely spiritual aspects of anxiety: "Deep within every human being there still lives the anxiety over the possibility of being alone in the world, forgotten by God, overlooked by the millions and millions in this enormous household."[21] Anxious despair over such general concerns as the existence of God, the inevitability of death, and the threat of meaninglessness, said Kierkegaard, was not pathological but an intrinsic and universal aspect of the human condition. Moreover, humans could be afraid not just of everything but also of nothing at all, because fundamental apprehensions about human existence underlay anxiety. In line with Christian tradition, Kierkegaard emphasized the connections of guilt, personal responsibility, and anxiety. In the face of an inevitable anxiety, individuals must choose either despairing or making a leap of faith to belief in God.[22]

It was difficult to incorporate such spiritual states of anxiousness into the somatic scheme. With the nineteenth century demanding a view of anxiety that was disease oriented, Kierkegaard's thought languished for a hundred years, until it became a major influence on existential philosophers and theologians in the mid-twentieth century.

NEURASTHENIA

As nervous moods were increasingly viewed as stemming from the brain and nervous system and so were framed within medical as opposed to moral or religious idioms, they needed to be described in medical terms. By midcentury, a variety of nervous conditions were listed in the records of medical practice. Sufferers from a range of ailments, physical and moral, such as headaches, phobias, insomnia, fatigue, and lack of purpose, clearly were not insane, but no distinct diagnostic terms captured their complaints.[23]

A new diagnosis, "neurasthenia," created a catch-all term reflecting popular theories of nervous energy, which healers and patients alike widely adopted. An American neurologist, George Beard (1839–1883), coined the term in 1869.[24] Beard was committed to developing a scientific basis for his specialty. Neurologists faced the dilemma of having to deal with outpatients who expressed an extremely diffuse array of woes that encompassed vague physical ailments, sexual problems, fatigue, and anxiety. The dominant theories they confronted indicated that such problems were signs of hysteria or hypochondriasis, that is, they were products of the patient's mind without a somatic basis. Yet, these theories were not congruent with what had become standard medical thinking at the time: first, real diseases must be somatic and, second, they were distinct entities that displayed common causes and prognoses.

Beard developed the neurasthenia diagnosis to indicate that the conditions he described were physical, not mental, states. A weakness of the nervous system underlay all of the various symptoms that fell into this category. While eighteenth-century physicians viewed the nerves as strong and energetic forces, by the time Beard was writing they were associated with excitability and consequent exhaustion. "Nervousness," Beard declared, "is a physical not a mental state, and its phenomena do not come from emotional excess or excitability or from organic disease but from nervous debility and irritability."[25] Too much stress overtaxed people's coping abilities and depleted their nervous energy, causing nervous exhaustion.

The neurasthenia diagnosis initially combined a protean mixture of various physiological and psychological conditions and so was ideally suited for the different manifestations of anxiousness and, indeed, of all the functional nervous diseases. Neurasthenics complained of numerous yet amorphous physical pains, including headache, stomachache, back pain, fatigue, skin rashes, insomnia, asthma, and poor general health. Both mental and somatic ailments were viewed as forms of nervous debility, not as psychological afflictions. Although symptoms of anxiety were not among the most prominent aspects of neurasthenia, irrational concerns, various phobias, and fears of contamination were explicitly part of this condition. The boundaries of neurasthenia and hysteria, another very general diagnosis that was commonly associated with female sufferers, were especially diffuse and difficult to discern. Nevertheless, the catch-all diagnosis of neurasthenia imposed an artificial coherence and precision on what had been seen as a disparate array of symptoms. Most importantly, the diagnosis provided patients with a somatic-sounding label for their symptoms that fit no existing category of organic disease.[26]

Beard echoed Cheyne's earlier notion that social forces created mental distress and disorder: the rapid increase of nervous conditions arose from the stresses of civilization. For Beard, neurasthenia could only have arisen in the nineteenth-century American environment that led people to become overly tense, worked up, on edge, and overwhelmed by stress. He noted, "No age, no country, and no form of civilization, not Greece, nor Rome, nor Spain, nor the Netherlands, in the days of their glory, possessed such maladies."[27] In particular, "steam power, the telegraph, the

periodical press, the sciences, and the mental activity of women"
accounted for the peculiarly American nature of the disease.[28] Although Beard thought that neurasthenia was a uniquely American
affliction, it also became an extremely popular diagnosis in Europe.
Paul Dubois, a French physician, observed, "The name of neurasthenia is on everybody's lips; it is the fashionable new disease."[29]
The preeminent French neurologist Jean-Martin Charcot asserted
that neurasthenia was a diagnosis that genuinely "correspond(ed)
to the reality of things."[30]

Although Beard stressed that social factors led to the ubiquity of neurasthenia, he also emphasized how heredity determined which particular persons succumbed to the pressures of
civilization. Individuals' lifestyles and experiences provoked their
inherited susceptibilities to nervous weakness. The biologically
predisposed could succumb to nervous collapse in several ways.
One stemmed from the hectic demands and constant stimuli of
modern life—mechanization, newspapers, the telegraph, scientific advancement. Another way was through decadent activities,
such as illicit sex, masturbation, or gambling. Mental and physical overexertion could also lead to the draining of nerve force and
consequent nervous breakdowns.[31] Like Cheyne, Beard believed
that the upper classes, artists, and white collar workers were especially prone to suffer from neurasthenia, thus associating the disease with social status, culture, and distinction. In line with Social
Darwinist theories that prevailed at the time, Beard believed that
the affluent and refined classes were more sensitive and thus more
prone to nervous exhaustion. Those who worked with their minds
were also particularly likely to succumb to neurasthenic conditions.[32] Because of their greater participation in civilization, men
were more vulnerable to neurasthenia, while women were more
likely to develop hysteria.

In fact, the supposed epidemic of neurasthenia more likely resulted from the appeal of the diagnosis to the newly emergent
middle class. "Patients," Edward Shorter writes, "found the notion of suffering from a physical disorder of the nerves far more
reassuring than learning that their problem was insanity."[33] Neur-

asthenia provided a somatic-sounding label that could protect patients from the idea that their problems were purely psychic. That Beard emphasized treatments employing drugs, injections, electricity, and the like and did not use psychological therapies helps account for the immense popularity the diagnosis enjoyed: it removed a stigma from people who suffered from what they and their physicians could believe was a genuine physical disease.

The most important contribution of the spread of the diagnosis of neurasthenia was to direct psychiatric attention to non-institutionalized patients. It was, in Beard's term, a "disease of the street," as opposed to the hospital.[34] The diagnosis broadened conceptions of what sorts of conditions could be legitimately treated as diseases in community practices. Moreover, it provided a non-stigmatizing label for those who were overcome by the pressures of modern civilization. The respectability of the diagnosis fanned its spread. "Within a decade of Beard's death in 1883," Charles Rosenberg notes, "the diagnosis of nervous exhaustion had become part of the office furniture of most physicians."[35]

The label of neurasthenia focused a previously scattered diagnostic scene into a single category that dominated descriptions of nervous conditions for the remainder of the century among both physicians and the general educated public. In retrospect, this vast category did not enhance the amorphous classificatory scene that existed at the time. It was, in essence, scientifically useless. As prominent English physician T. Clifford Allbutt (1836–1925) noted, "The so-called diseases of the nervous system [were] a vast, vague, and most heterogeneous body, two-thirds of which may not primarily consist of diseases of nervous matter at all."[36] Nevertheless, neurasthenia became the major diagnosis of nervous afflictions in the late nineteenth century because it seemed to satisfy the needs of the medical profession for a distinct diagnosis of nervous complaints and the desires of patients to receive a physical label for their nervous ailments. The diagnosis, however, would succumb to the growing demands for the specification of distinct anxiety states.

THE GROWING SPECIFICATION OF ANXIETY

Since the Hippocratics, physicians had viewed diseases holistically, highlighting their disruption of the balance of forces within individuals and between individuals and their environments. Even when eighteenth-century physicians turned to the nervous system as the source of disease, they emphasized how local lesions produced systemic effects. Notions of disease specificity had no place in their broad, multicausal views of illness.[37]

By the beginning of the nineteenth century, however, notions of specificity grounded in anatomy, cell pathology, and microbiology were replacing holistic conceptions of mind and body within general medicine. The discovery of bacteria and viruses began to transform theories of disease causation. Echoing Sydenham's earlier conceptions, medical thinking began associating diseases with specific mechanisms that were unrelated to the particularities of the individual people who harbored them. Each distinct disease was presumed to have a characteristic cause, prognosis, and underlying somatic pathology.[38] Moreover, the specific sites of drug reactions were becoming known, so drugs were increasingly thought to have local as opposed to general effects.

Psychiatry did not yet participate in these trends. Indeed, the new conception of disease was diametrically opposite to psychiatry's conceptions of nervous disorders at the beginning of the century. At this time, although the nervous system had replaced the humors as the locus of nervous disorders, psychiatrists still viewed these conditions holistically, as inseparable from general well-being. While physicians had discarded humoral conceptions, they maintained virtually intact the Greek belief that, in the case of mental illness, temperament and other general factors, including heredity, constitution, environment, and trauma, were predisposing influences. For psychiatrists, Charles Rosenberg explains, disease was still "conceived of not as a set of specific entities, each with a characteristic and generally predictable course and underlying mechanism, but as a physiological state of the individual patient; all causation was multicausal."[39] Mind and body were interconnected: bodily disturbances led to mental ones, as well as the converse. From this perspective, health and disease remained inextricably tied to the relationship between individuals and their environments. No standardized diagnostic system for mental illnesses existed in part because local lesions had general, not just specific, effects.[40]

For the first half of the nineteenth century, psychiatric classifications didn't systematically distinguish anxious from melancholic, hypochondriacal, and hysterical conditions. Most diagnosticians continued to link anxiety and depression. The Belgian psychiatrist Joseph Guislain (1797–1860) clearly described a type of generalized anxiety as a form of melancholia. "There is a whole series of melancholias in which the patient is dominated by vague worries. He feels ominous premonitions. He doesn't feel well anywhere; terrible misfortune seems to threaten him; he's fearful of everything, afraid of everything."[41] Nor were specific therapies indicated to treat particular nervous complaints. Most psychiatrists and physicians were nondogmatic, employing whatever combination of somatic and psychological responses seemed most appropriate in each individual case. Typical treatment regimens included diet, exercise, fresh air, rest, and drugs, a combination that would not have been unfamiliar to Hippocratic physicians.

The latter half of the nineteenth century saw dramatic changes in the understanding of anxiety, as increasingly differentiated diagnoses upended the two thousand–year tradition of holistic thinking rooted in interactions between mind, behavior, and environment.[42] Although most symptoms of anxiety continued to be submerged in gross categories, especially the catch-all neurasthenia, toward the end of the century a number of independent anxiety conditions emerged. While a few earlier publications exist, most of the first authoritative articles on specific anxious conditions appeared in the 1870s, and there was a flurry of them. Benedikt and Westphal described panic attacks and agoraphobia, respectively. A French neurologist, Edouard Brissaud (1852–1909), described a type of panic attack (which he called "paroxystic anxiety") in which patients believed they were about to die despite having no physical symptoms. Around the same time, Hungarian ear-nose-throat specialist Maurice Krishaber (1836–1883) described a new condition that he labeled "cerebro-cardiac neuropathia." It was very similar to what later came to be called "anxiety neurosis," featuring sudden attacks of intense anxiety, pounding heart, and dizziness that weren't related to any definite object.[43] Obsessive-compulsive conditions also began to be separated from other types of anxiety, with some observers focusing on the disturbed thoughts they represented. Later, in 1901, a French psychiatrist, Paul Hartenberg (1871–1949), provided the first medical description of what is now known as "social phobia." His book *Les timides et la timidité* emphasized how the presence of other humans precipitated in some people feelings of intense anguish, sweating, and heart palpitations.[44] Such distinct conditions were not easy to fit into the capacious categories of neurasthenia and hysteria. A French psychologist, Theodule-Armand Ribot (1839–1916), criticized the tendency to coin distinct terms for each type of phobia and anxiety, but he did distinguish generalized anxiety states from phobias. By the last decades of the century, specificity was clearly influencing conceptions of anxiety.[45]

In contrast to the development of ever more specific descriptions of anxiety conditions, German psychiatrist Emil Kraepelin (1856–1926), the most prominent diagnostician of the late nineteenth and early twentieth centuries, gave short shrift to anxiety as a freestanding disease syndrome. Kraepelin was a hospital-based psychiatrist who was particularly concerned with distinguishing the psychotic conditions that afflicted most of his inpatients by separating their differing prognoses.

Kraepelin was well aware of the centrality of anxious emotions. He noted the omnipresence of natural anxiety: "Even in normal individuals it affects sympathetically the entire mental and physical condition, being accompanied by precordial oppression, palpitation, paleness, increased respiration and tremor, and sometimes by perspiration and an increased tendency to urinate and defecate." Moreover, among the mentally ill, he wrote, "Fear is by far the most important persistent emotion encountered in morbid conditions."[46] However, Kraepelin studied anxious excitement and anxious tension in other mental disorders, especially manic-depression. He also discussed *angst* in his famous text, *Clinical Psychiatry*, but he emphasized its close connection to melancholia in the category "anxious melancholia." Kraepelin's neglect of anxiety as an independent condition starkly contrasts with his seminal influence on the classification of schizophrenia and mood disorders. Given his preoccupation with psychotic disorders, it is not surprising that Kraepelin stood apart from the growing specification of anxiety disorders and instead continued to connect anxiety to broader psychotic states.

By the century's end, such expansive categories as neurasthenia and hysteria began to give way to specificity. In addition, classifications had come to clearly distinguish anxiety from psychotic conditions, which were considered severe derangements of thought and reason. By the end of World War I, use of the diagnoses of neurasthenia and hysteria had run their course. The somatic nature of neurasthenia, which had originally created its popularity during the nineteenth century, led to its demise in the following century when the class of neuroses moved from the physical realm to the psychological. By the early twentieth century, psychiatrists were arguing that neurasthenia and related conditions should be treated within the newly emerging psychiatric sector, rather than in general medicine.[47] Moreover, nothing whatsoever about brain pathology had been discovered that could account for nervous afflictions, nor had the emphasis on brain pathology led to any therapeutic progress. The time was ripe for a unification of a distinct class of anxiety disorders with a unique identity.

TOWARD THE UNIFICATION OF ANXIETY DISORDERS

As the nineteenth century came to an end, diagnosticians were carving anxiety into an increasing number of specific conditions. Notwithstanding Morel's efforts at integration, the variety of distinct conditions lacked any unifying underlying principle. Pierre Janet (1859–1947), a French neurologist and psychiatrist, developed a more general conception of anxiety in his *Les obsessions et la psychasthenie* (1903). Janet's goal was to explain problems of everyday life, not just serious psychiatric disorders, and he found the overly broad concept of neurasthenia virtually useless for this purpose. He viewed anxiety as a vague expression of respiratory and cardiac complaints that often underlay other emotions, such as anger, fear, and love. However, he thought that these symptoms were primarily expressions of psychological, not somatic, problems, and considered anxiety to be "the most elementary of the mental functions."[48]

Janet strove to unify the various specific entities that diagnosticians had proposed, partly on an etiological basis. He divided the neuroses into two major classes, hysteria and psychasthenia. The former condition featured disturbances of consciousness, sensation, and movement, which were caused by traumas that had been subconsciously repressed and replaced by pathological symptoms. Psychasthenia referred to a collection of symptoms of anxiety, phobia, obsession, fatigue, and depression. While hysterical conditions emerged from unconscious processes, psychasthenic ones affected the conscious mind. Fears—including fears of death, anticipated catastrophes, or of taking action of various sorts—often led to psychasthenic conditions. In contrast to his differentiation of anxious and hysterical symptoms, Janet emphasized a close link between anxiety and depression: "la tristesse peut prendre la forme de melancholie, d'anxiété, d'angoisse morale (the sadness may take the form of melancholy, anxiety, mental anguish)."[49]

Janet divided the general class of psychasthenia into three categories: obsessive thoughts, irresistible compulsions, and visceral states, the last of which included generalized anxiety, panic, and phobias. These states were interrelated and often merged. Moreover, if one state was suppressed, it could reappear in another form. All types of psychasthenia were disturbances of psychological functioning that resulted in feelings of incompleteness, depersonalization, derealization, and exhaustion. Although hereditary factors could predispose people to these conditions, the symptoms themselves were psychological and unique to the individual. This differentiated these factors from the physical symptoms associated with neurasthenia. Janet also developed a system of psychotherapy, although he acknowledged that therapy would often not be successful.

Janet differentiated his views from those of Freud, who he believed overemphasized the importance of sexual repression. In another difference between them, Janet felt that individuals with inherited predispositions to neuroses had less than normal quantities of nervous energy, not excesses of undischarged energy, as Freud believed. Janet's theories were influential at the end of the nineteenth and beginning of the twentieth century, but, unlike Freud, he did not found a school that would carry on his work, so it quickly slipped into obscurity.[50] Nevertheless, Janet's was the most impressive effort, aside from Freud's, to develop a comprehensive nosology of anxiety conditions. By the century's end, a new psychological view of anxiety had emerged to compete with the model grounded in the central nervous system.[51]

CONCLUSION

Although the medical labels attached to nervous conditions during the nineteenth century were inexact and reflected uncertainty over whether anxiety was primarily a somatic or a psychological problem, over the course of the century mental troubles were transformed from inevitable human afflictions to diseases. "The nineteenth-century age of science," historian Theodore Zeldin observes, "turned the old-fashioned passions into sources of illness instead of accepting them as the inevitable condition of humanity. When people worried, therefore, or were deeply depressed, they now expected a medical name for their condition and they got it."[52]

Likewise, the treatment of nervousness and other non-insane conditions had become the province of a new profession, nerve doctors. Indeed, by 1894, D. Hack Tuke could introduce a lecture to the Neurological Society with a quote from Hughlings Jackson: "The slightest departures from the standard of mental health are to be studied, and not only the cases of patients who require to be kept in asylums. Indeed, the slightest cases are the more important in a scientific investigation."[53] Anxiety had emerged as a problem related to the malaise of the middle- and upper-middle-class groups who would become the clients of a new sort of psychiatric practitioner. The century's developments culminated in the emergence of Sigmund Freud's theories and treatments, which both came to dominate psychiatry and to embody an entire cultural epoch.

CHAPTER FIVE

The Freudian Revolution

FREUD'S ACHIEVEMENTS

The emergence of Sigmund Freud as the dominant psychiatric theorist of the first half of the twentieth century was a central turning point in the history of anxiety. Previous diagnosticians usually viewed anxiety as an aspect of a broader entity, be it Hippocratic melancholia, Cheyne's nervous disorders, or Beard's neurasthenia; they did not distinguish anxiety-related conditions from depression, psychosomatic problems, or hysteria. When neurologists and psychiatrists began to recognize particular anxiety disorders, such as agoraphobia and panic attacks, in the nineteenth century, they did not connect them to a broader class of anxiety conditions. Freud's early works unified a variety of somatic and psychic expressions that previously had been considered distinct under the broad umbrella of anxiety conditions. His later works then moved anxiety to the core of psychiatric nosology.[1]

Despite the scorn that psychiatry now heaps upon his views, Freud was a major precursor to current thinking about anxiety. His work is generally acknowledged as the inspiration behind the first modern psychiatric manuals, the psychodynamically based *DSM-I* and *DSM-II*, which emphasized the dynamic relationships between unconscious motivations and conscious actions. In fact, it also foreshadowed the symptom-based approach of the subsequent manuals, *DSM-III* through the *DSM-5*. Paradoxically, Freud's early writings on anxiety that distinguish a variety of anxiety conditions anticipate the "neo-Kraepelinian" approach to psychiatric diagnosis. In current *DSM*-style, Freud used symptoms to identify distinct anxiety syndromes, a process that not only split anxiety from the broader conceptions encompassed in neurasthenia and hysteria but also identified the major anxiety conditions that persist in present-day psychiatry.

Freud's interest in anxiety disorders was apparent in his earliest writings and his deep concern with them never wavered throughout his career, despite substantial changes in his conceptions of their nature and origin. Although his particular etiological claims are no longer influential or even credible, Freud helped forge the scientific study of anxiety and build an enduring foundation for outpatient mental health treatment.

FREUD'S BACKGROUND

Sigmund Freud (1856–1939) was a nonobservant Jew who lived nearly his whole life in Vienna. For most of his life, the city was the capital of the sprawling Hapsburg Empire and the center of a vibrant cosmopolitan culture. It was also the medical hub of the world at the turn of the nineteenth century. Freud initially trained as a neurologist and spent ten years studying brain processes in a variety of species. He began his psychiatric career as an acolyte of French neurologist Jean-Martin Charcot (1825–1893) after studying with Charcot for a few months in 1885–1886. In accord with French medicine at the time, Charcot fervently believed in the hereditary and organic grounds of hysteria. Freud's early development, therefore, occurred in an environment steeped in hereditarianism and biological studies of mental illness.

In the embryonic stage of psychoanalysis, Freud teamed with Austrian physician Josef Breuer (1842–1925) to study hysteria. Far more than either Beard or Charcot, Freud and Breuer emphasized how intrapsychic factors led to neuroses. In particular, they highlighted the importance of suppressed memories of traumas in generating neurotic conditions and of hypnosis as a technique for recovering them. Freud, however, quickly repudiated his initial view that hysterical symptoms symbolically reproduced repressed traumatic experiences, and he came to stress how unconscious memories more likely resulted from patients' fantasies than from actual traumas. This was a "decisive turning point in psychoanalysis" because it moved Freud to study unconscious intrapsychic dynamics and their ambiguous relationship to tangible experiences.[2]

FREUD'S EARLY WRITINGS ON ANXIETY

Although psychiatrists had started to pay attention to panic, phobias, and obsessions by the time Freud began writing in the early 1890s, the study of anxiety remained in the shadow of neurasthenia and hysteria. Most symptoms of what were to become the various psychoneuroses were seen either as distinct syndromes or as manifestations of the capacious neurasthenic diagnosis, which was so broad that it encompassed virtually every sort of psychological symptom that general physicians and neurologists were likely to encounter. Freud was well-versed in Beard's writings, regarding them as too vague and overinclusive to be useful.

During the last decade of the nineteenth century, Freud came to reject Charcot's theory of hysteria and began to develop a unique psychological theory of the mind. At the core of his initial theory of hysteria was the idea that psychic repression blocked the direct discharge of sexual energy and channeled it into substitutive physical and psychological symptoms.[3] This repression of unconscious sexual memories and drives—rather than inherited predispositions—caused most hysterical conditions, he thought. Moreover, Freud felt that Charcot's emphasis on the hysterical conversion of psychic to somatic symptoms could not explain the psychic manifestations that were prominent in anxiety, phobic, and obsessive conditions.

Although anxiety disorders were at the heart of many of Freud's earliest papers, written between 1894 and 1897, commentators have generally ignored those writings. This neglect perhaps stems from the fact that they reflected the biological and diagnostic emphases that characterized late nineteenth-century psychiatry more than they did the revolutionary doctrine of psychoanalysis that Freud subsequently became identified with. These early works have been overshadowed by *Studies on Hysteria* (written jointly with Joseph Breuer in 1895) and *The Interpretation of Dreams* (1900), which were the foundational documents for psychoanalytic practice and theory respectively. His early writings on anxiety, however, have had more lasting influence on psychiatric diagnosis. The various specific entities that they delineated—general anxiety, phobias, obsessions, and panic—have largely persisted.

In 1894 Freud published a pathbreaking article, "The Justification for Detaching from Neurasthenia a Particular Syndrome: The Anxiety-Neurosis." From the morass of neurasthenia, Freud described a distinct syndrome with symptoms both closely related to each other and different from other symptoms of neurasthenia. As he put it,

> The symptoms of [anxiety neurosis] are clinically much more closely related to one another than to those of neurasthenia proper (that is, they frequently appear together and replace each other during the course of the illness), while the aetiology and mechanism of this neurosis are essentially different from what remains of true neurasthenia after this subtraction has been made of it.[4]

Freud thus created a separate class for anxiety disorders.

Freud identified a common force that he saw as underlying the anxious conditions of generalized anxiety, anxiety attacks, phobias, and obsessions. This force was

> something which is capable of increase, decrease, displacement and discharge, and which extends itself over the memory-traces of an idea like an electric charge over the surface of the body. We can apply this hypothesis . . . in the same sense as the physicist employs the conception of a fluid electric current. For the present it is justified by its utility in correlating and explaining diverse psychical conditions.[5]

The "something" Freud referred to was anxiety. Freud grouped together a range of syndromes that had been viewed as diverse conditions and identified them as different manifestations of the same underlying process. A range of disturbances—gastrointestinal upset, heart and breathing problems, psychic obsessions, panic, general unease, behavioral compulsions or inhibitions—were all varying expressions of a common anxious energy force.

Freud distinguished the anxiety neuroses from both neurasthenia and hysteria. Neurasthenia, he asserted, resulted from unsatisfied sexual needs. Like other physicians at the end of the nineteenth century, he viewed masturbation, in particular, as a major source of neurasthenia. While neurasthenia resulted from inadequate forms of sexual release, anxiety neuroses, declared Freud, stemmed from the accumulation of more sexual excitation than people could psychically assimilate. Anxiety neuroses were thus especially common among sexually frustrated practitioners of *coitus interruptus*, sexual abstinence, and the like.[6]

He also distinguished the anxiety neuroses from hysteria. Anxiety conditions came from current sexual problems that were "not derived from any psychical source." In contrast, hysterical conditions were rooted in psychological repressions of earlier real or imagined traumatic memories, so hysterical patients suffered "principally from reminiscences."[7] Hysterics escaped traumatic memories by using defenses that transformed unacceptable ideas into more tolerable bodily expressions. These memories ultimately were of a sexual nature, although repression often converted them into different, usually somatic, expressions.

Thus, Freud's view grounded hysterical suffering in the psychological realm and anxious conditions in somatic processes. He further stated, "anxiety-neurosis is actually the somatic counterpart of hysteria."[8] Although Freud himself did not think that psychological conditions were any less involuntary than physical ones, regarding anxiety as a disturbance of organic functioning was more palatable at a time when psychogenic conditions were not yet legitimated. One result of his view was that anxious conditions became more socially acceptable than hysterical ones.

Freud not only severed organically based anxiety conditions from neurasthenia and hysteria but also delineated several anxiety conditions. The core anxiety symptom was "anxious expectation," a free-floating state of nervousness and apprehensiveness, which is similar to the diagnosis of Generalized Anxiety Disorder (GAD) that emerged in *DSM-III*.[9] Freud provided this example:

> A woman who suffers from anxious expectation will imagine every time her husband coughs, when he has a cold, that he is going to have influenzal pneumonia, and will at once see his funeral in her mind's eye. If when she is coming towards the house she sees two people standing by her front door, she cannot avoid the thought that one of her children has fallen out the window; if the bell rings, then someone is bringing news of a death, and so on.[10]

Anxious expectation, Freud noted, could attach itself to any suitable ideational content.

Numerous commentators from St. Augustine through Kierkegaard had described anxiety conditions similar to generalized anxiety as being omnipresent aspects of the human condition. Freud's conception, however, moved generalized anxiety from the religious and philosophical spheres to the domain of medical practice. It was a significant step toward the medicalization of anxiety that had no specific content.

Freud also described "anxiety attacks," which closely resembled what are now called "panic attacks." These featured anxiety that suddenly burst into consciousness along with some disturbance of respiratory, cardiac, vascular, and/or glandular function.[11] In addition, he carefully separated the symptoms of phobic and obsessive conditions, describing two types of phobias. One had to do with fears of snakes, thunderstorms, darkness, or vermin, and thus exaggerations of instinctual aversions that took to extremes reactions shared by most people. The second dealt with fears of objects and situations such as open spaces or writing that were not usually fearful. He believed that phobias, unlike hysteria, did not originate in repressed ideas and so were not amenable to psychoanalysis.[12]

Obsessions—intrusive and unwanted ideas that patients could not excise from their consciousness and needed to resist—Freud placed in the general category of anxiety disorders. He distinguished obsessions from phobias: obsessions were related to fixations on certain thoughts and rituals that forced themselves on patients as opposed to intense phobic fears of particular objects or situations.[13] An obsession substituted a more tolerable idea for a painful idea connected to an unacceptable sexual experience that the individual wanted to forget. Sufferers recognized that the obsession was not rational, but they felt powerless to resist the force of the undesirable intrusion.

Freud's early work unified under the umbrella of "anxiety" a variety of previously distinct syndromes or symptoms that had been associated with broader conditions. His split of anxiety from neurasthenia, hysteria, and other states, his creation of a distinct class of anxiety conditions, and his delineation of distinct anxiety states both fulfilled the movement of growing specificity that marked nineteenth-century conceptions of anxiety and anticipated the *DSM-III* diagnostic revolution of 1980.

FREUD'S INITIAL ETIOLOGY OF ANXIETY DISORDERS

Freud not only distinguished the particular symptoms of the major anxiety conditions but also provided an etiological basis for separating these conditions from one another and from other types of neuroses. A physiological view underlay Freud's initial etiology of anxiety. In the 1890s, the basic principle of neurology was that psychic experiences reflected physical and chemical processes. Neurologists accepted the laws of the conservation and transformation of energy developed in physics and adopted them to explain the operation of the brain.[14] Freud used these principles to explain how nervous energy could be transformed into a variety of psychic, behavioral, and somatic manifestations that were representations of the same underlying force.

Freud's most original (and most criticized) contribution to the etiology of anxiety was to connect the conservation and transformation of anxious energy specifically to sexual energy. For him, psychological well-being depended on an adequate discharge of sexual excitation. In this hydraulic model, drives that were blocked in one way would find expression through some other way. Freud relentlessly pressed for the sexual basis of all the neuroses: frustrated sexual energy would invariably be transformed into neurotic symptoms. Anxiety was a somatic force created by dammed-up libido.

Definitions of individual neuroses were often overdetermined, Freud stressed: while pure cases were thought to have specific causes, in reality many cases involved numerous causative factors. Regardless of the number of factors, Freud consistently emphasized the primacy of sexual causes, insisting that sexual factors underlay every case of neurosis. He distinguished various conditions from one another by linking each to particular kinds of sexual frustration. Freud's initial writings never ceased to emphasize the pervasive power of sexual energy in shaping neurotic illnesses and to stress how disturbances in present or past sexual life invariably caused a neurosis:

> Each of the major neuroses has as its immediate cause a special disturbance of the nervous economy, and that these pathological functional changes *betray, as their common source, the sexual life of the person concerned, either a disturbance of his present sexual life or important events in his past life.*[15]

The symptoms of all neuroses were products of sexual activity, whether current or past, real or fantasized. Other forces, such as heredity or temperament, might be predisposing, contributing, or supplementary factors, but only sexuality had the power to cause a neurosis in the absence of any other forces. "No neurosis," Freud summarized, "is possible with a normal *vita sexualis.*"[16]

This view of anxiety reflected the biological emphasis that dominated nineteenth-century thinking. Freud's basic etiological division was between psychoneuroses and actual neuroses. Psychoneuroses, which included hysteria and obsessions, stemmed from repressed memories of real or imagined sexual traumas in childhood, he thought, while the actual neuroses of anxiety and

neurasthenia resulted from current sexual disturbances and reflected problems in expressing somatic energy.[17] This difference in
origin put anxiety outside the major thrust of his theories of that
time, which emphasized the intrapsychic causes of the psychoneuroses. Indeed, Freud suggested that ultimately anxiety neuroses
would no longer be considered neurotic disturbances because, at
bottom, they reflected metabolic disturbances of sexuality.[18]

Anxiety also required different treatments from those effective for the psychoneuroses, Freud claimed. Psychoanalysis was
best able to deal with psychoneurotic conditions like hysteria and
obsessions, which resulted from repressed childhood memories,
while the actual neuroses were less suitable for analysis because of
their somatic causes in aroused but ungratified sexual excitation.
The therapeutic task in treating actual neuroses was to convince
patients to engage in normal sexual relations and to refrain from
practices such as *coitus interruptus* and masturbation.[19]

The particular form a neurosis took was determined by the way
in which the individual dealt with sexual excitation. Sexual energy
could be repressed (hysteria), discharged inadequately (neurasthenia), or not allowed discharge (anxiety). The fact that each style of
neurosis had a separate cause did not prevent their co-occurrence.
Freud noted the common co-presence of anxiety conditions with
other conditions, such as neurasthenia, hysteria, and melancholia.
Indeed, a single individual typically suffered from more than one
condition because of the common co-presence of the specific factors related to each condition, for example, among masturbators
prone to neurasthenia who become abstinent and therefore also
liable to develop anxiety.[20]

In contrast to Freud's initial way of classifying anxiety, which
largely persists in current psychiatric manuals, history has not been
kind to his etiological theory based on sexual energy. The failure
of Freud's early theory of the etiology of anxiety seems largely
due to the degree to which it reflected his particular cultural circumstances rather than universal somatic and psychic processes.
Viennese culture was obsessed with thoughts of sex, which often
became manifest through repressed and distorted expressions. Its

cultural norms featured an especially powerful prohibition against masturbation; many physicians emphasized its horrific consequences, among which they included insanity, premature death, and vulnerability to many diseases.[21]

Freud's writings illustrate this preoccupation. They were intensely concerned with the nefarious impact of sexual repression on resultant psychopathology, in particular how repressed sexual energy brought about anxiety conditions. That he pinpointed the source of anxiety in the accumulation of too much sexual energy and its corresponding cure in more adequate sexual discharge had an obvious appeal in a milieu that negatively sanctioned all but the most limited forms of sexuality. Better sex as a therapeutic technique tied anxiety neuroses to fashionable ideas about sexual liberation. Focusing on sexual repression as the source of neuroses and on expressing sexual energy as a means to psychic health inspired those who opposed conventional morality and sought a sexually freer society. It is not surprising that many of Freud's first followers were bohemians, artists, and writers. Freud himself underwent a long and painful period of sexual repression during his extended engagement to his eventual wife, Martha. They did have five children, but they ceased sexual relations for the last several decades of their lives.[22]

The genius of Freud's early works on anxiety resides not in his etiological theory but in his differentiation of the major types of anxiety—generalized, panic, phobias, and obsessions—and their unification in a single diagnostic class. Moreover, he distinguished these conditions from the more prominent extant entities of hysteria and neurasthenia. His subsequent writings developed different etiological and diagnostic conceptions that placed far less emphasis on specific diagnoses and on an exclusively sexual and organic etiology of the anxiety neuroses. They also elevated the standing of anxiety, identifying it as the force responsible not only for anxiety conditions per se but also for all of the psychoneuroses.

THE TRANSFORMATION OF FREUD'S CONCEPT
OF ANXIETY

The anxiety neuroses were central to Freud's early writings, reflecting his efforts to tie the neuroses to sexual concerns. At the beginning of the new century, however, Freud began to reject the idea that mental disorders emerged from physiological forces. One of his disciples, Wilhelm Stekel (1868–1940), had disputed Freud's view that many anxiety disorders were caused by current sexual inadequacies and instead claimed that they arose from psychic conflicts. Moreover, Stekel said he had cured such cases by bringing unconscious conflicts into consciousness. Although Freud publicly repudiated Stekel's interpretations, by 1909 he himself had partially acknowledged the theory by incorporating phobias into the class of psychically caused conditions.[23]

Freud gradually developed a more psychological approach to all of the neuroses, especially emphasizing processes of repression. One of Freud's most famous cases, Little Hans (1909), illustrates both a turning away from his notion of dammed-up libido as the primary source of anxiety and a movement toward awareness of the power and importance of psychic repression. Hans was a five-year-old boy who developed a serious phobia about horses. Shortly after seeing a horse fall to the ground, the boy became intensely anxious, clung to his mother, and was terrified of going onto the street because of a fear that a horse would bite him. Freud interpreted Hans's symptoms as indicating the repressed castration anxiety that typified the Oedipal complex rather than as the undischarged sexual energy that characterized his initial theories of anxiety. He viewed Hans's phobic symptoms as mechanisms that served to relieve the anxiety and tension that his unconscious conflict with his father created. Repression forced Hans's unacceptable thoughts into his unconscious, blocked their overt discharge, and transformed them into substitutive symptoms.

According to Freud, Hans wanted to kill his father in order to sleep with his mother but instead transferred his murderous instincts toward his father onto horses. This displacement supposedly allowed Hans to remain unaware of his rage toward his father and so maintain his love for him. The horse phobia also let Hans avoid what he feared by simply not going outdoors. The true object of his fear, his father, could not be so easily avoided. Repression had displaced undischarged sexual energy as the cause of anxiety.[24]

World War I led to another turning point in Freud's thoughts about anxiety. The psychological impacts of the war led him to revise a number of aspects of his theory. Thousands of combatants developed what came to be called "shell shock," which typically involved paralysis of limbs, hysterical loss of functions of sight, hearing, or speech, and/or various tics and fits. This flew in the face of Freud's theory that such symptoms were caused by repressed psychic memories of past events. Freud could not ignore the powerful impact of external and contemporaneous traumas on the development of anxiety and other conditions. The war-time disorders, which were not psychic in nature but stemmed from actual events, showed that sexual conflicts need not be the sole source of neuroses. These widespread traumas challenged the body of work that Freud had developed over the previous two decades and paved the way for a fundamental revision of his theory.

Inhibitions, Symptoms, and Anxiety (1926) summarized Freud's later writings on anxiety. The book was written largely as a response to Freud's former disciple, Otto Rank (1884–1939), who posited that birth was the primal traumatic experience. Rank believed that being born provided the template for future danger situations and the development of subsequent anxiety. Freud agreed that birth trauma was a forerunner of subsequent neuroses but did not believe that it was their exclusive prototype. He began to move the key source of anxiety from somatic and intrapsychic factors to the ways in which the ego reacted to dangerous situations, especially separations from mothers.[25]

Freud's conception of anxiety also broadened, from its intimate connection with sexual energy toward a more general unpleasant response. He came to reject the hydraulic model of sexual energy that had underlain his initial conceptions of anxiety and to see anxiety no longer as transformed libido but as a danger signal triggered by trauma. In addition, he recognized that anxiety could stem from the anticipation of future dangers. He began to place a greater emphasis on indefinite anxiety (*angst*) that lacked any object. In his evolving view, anxiety was no longer an exclusive product of sexual repression but could emerge whenever a potent stimulus overwhelmed a person. Its intensity varied according to the ratio between external demands and the person's ability to handle them: any dangerous situation that was more powerful than the person's self-protective resources could lead to feelings of helplessness and, consequently, anxiety.[26]

Freud did not completely abandon his initial theory of anxiety. While he came to place more emphasis on external traumas, he also declared that actual traumatic events in themselves rarely led to neuroses: "it would seem highly improbable that a neurosis could come into being merely because of the objective presence of danger, without any participation of the deeper levels of the mental apparatus."[27] Traumas evoked conflicts that were already present, so latent predispositions and external events jointly created disorders. People who experienced traumas and consequently developed neuroses also had deeper, unconscious factors that contributed to their symptoms.

Freud's later works also attempted to distinguish normal from pathological forms of anxiety. An anxious condition was normal or pathological depending in part on the life stage when it developed. Freud concluded that each stage in personal development had prototypical, normal sources of anxiety. The primary source of anxiousness among infants was the expectation of loss of the baby's mother. Fears of being left alone, of being in the dark, and of strangers were the most common manifestations of early anxiety. During the "phallic phase," boys especially feared the castrating power of their fathers while object loss was the primary source of anxiety among girls. Older children were afflicted by anxiety inspired by fears of punishment from the moralistic superego.[28]

Disordered states were regressions to earlier stages of development:

> We consider it entirely normal that a little girl should weep bit-
> terly at the age of four if her doll is broken, at the age of six if
> her teacher reprimands her, at the age of sixteen if her sweetheart
> neglects her, at the age of twenty-five, perhaps, if she buries her
> child. . . . We should be rather surprised, in fact, if this girl, af-
> ter she had become a wife and mother, should weep over some
> knickknack getting broken. Yet, this is how neurotics behave.[29]

Normal people, in Freud's view, outgrew their childhood fears, while neurotics still behaved as if old danger situations still existed. For example, fears of being alone and of the dark that would be normal in young children were symptoms of anxiety neuroses in adults. Freud confidently declared, "In not a single adult neurotic do the indications of a childhood neurosis fail of occurrence."[30]

Another distinction between natural and neurotic anxiety conditions was the degree of uncertainty in the anxiety-provoking situation: "A *real* danger is a danger which we know, a true anxiety the anxiety in regard to such a known danger. Neurotic anxiety is anxiety in regard to a danger which we do not know."[31] Normal anxiety served as a useful warning of an impending state of danger and could possibly help avert the danger from occurring at all. Neurotic anxiety, in contrast, arose when the amount of indefinite anxiety surpassed the ego's coping capacities and forced the neurotic to employ irrational psychic defenses. No clear line, however, separated normal from abnormal anxiety; they gradually shaded into each other.

THE CONSEQUENCES OF FREUD'S LATER THEORY OF ANXIETY

The most important theoretical shift revealed in *Inhibitions, Symptoms, and Anxiety* was Freud's reversal of his earlier assertion that repression transformed unsatisfied libido into anxiety. Instead, he wrote, "It was anxiety which produced repression and not, as I formerly believed, repression which produced anxiety."[32] This conclusion not only inverted the causal sequence of repression and anxiety but also signaled a fundamental alteration of Freud's initial view of anxiety as a somatic force. In his final opinion, anxiety served as a signal of some threat, warning of impending danger and activating defense mechanisms. In statements reminiscent of Aristotle's observations about dangerous situations, Freud described how the ego judges that a not-yet traumatic situation has the potential to become so:

> The conclusion we have come to, then, is this. Anxiety is a reaction to a situation of danger. It is obviated by the ego's doing something to avoid that situation or to withdraw from it. It might be said that symptoms are created so as to avoid the generating of anxiety. But this does not go deep enough. It would be truer to say that symptoms are created so as to avoid a *danger-situation* whose presence has been signalled by the generation of anxiety.[33]

While real anxiety was a response to some known external danger, neurotic anxiety was a reaction to some unconscious threat. Such internal threats included sexual ones but could also emerge from helplessness related to separations or threats to self-preservation.

The idea of anxiety as arising from some inner danger was unique at the time, and it had lasting influence on the field of psychiatry.[34] The emphasis of Freud's theory in general changed from the study of drives and instincts to the study of the ego and its defenses. "As a consequence of these new theories," psychiatrist and historian Henri Ellenberger asserts, "the ego was now in the limelight of psychoanalysis, especially as the site of anxiety: reality anxiety, that is, fear caused by reality, drive anxiety from pressures from the id and guilt anxiety resulting from the pressures of the superego."[35] Psychoanalytic treatment as well turned from the recovery of repressed memories toward the strengthening of the ego as the embattled mediator of pressures from instinctual drives and the moralistic superego.

As a result of Freud's reversal of the causal sequence of anxiety and repression, anxiety became the foundational condition of the neuroses. Freud had initially viewed repression as responsible for hysteria, sexual problems, minor depression, and a host of other neuroses. If, instead, anxiety was the primary force leading to repression, it also must be the principal reason for the emergence of the other psychoneuroses. (Freud associated serious depression with melancholia, so it remained outside the category of the newly fashionable neuroses.) Freud's later work enshrined anxiety as the foundation of all of the psychoneuroses and placed it on a pedestal that was far higher than in all previous psychiatric history.

Freud's later theory was also far less diagnostically specific than his initial discussions of anxiety. He now regarded anxiety as a general signal of danger that prompted the ego to employ various defense mechanisms connected with repression. Particular symptoms were not so much indicators of a distinct type of neurosis as overt expressions of underlying psychodynamic processes. The life history of the individual determined what specific appearance the symptoms of anxiety would take. Anxiety itself, rather than its various expressions, became "the fundamental phenomenon and main problem of neurosis."[36]

The expanded concept of anxiety that emerged in Freud's later work in certain ways turned psychiatric nosology back toward the more general views that had dominated its classifications until the mid-nineteenth century. The particular symptoms were less important than the mechanisms that underlay them. The chief difference was that anxiety, rather than melancholia, nerves, or neurasthenia, became the unifying element that explained the variety of diverse expressions. Ultimately, the downplaying of diagnostic specificity in Freud's later works, which ran against the dominant trends of twentieth-century medicine, became a major reason for the subsequent demise of psychoanalysis.

Freud's later view of anxiety was especially consequential for psychiatry, because it allowed the profession to gain dominion over a very common condition and differentiated psychiatrists from both hospital-based alienists and brain-focused neurologists. On the one hand, psychoanalysis severed the tie between psychiatry and the asylum by focusing on conditions that did not require hospitalization. On the other hand, psychologically based illnesses set psychiatry apart from neurology. Anxiety, as well as the sexual and interpersonal problems caused by underlying anxiety, became the bread and butter conditions that patients brought to the out-patient psychiatric practices established at the beginning of the twentieth century. This separation from neurology, however, also placed psychiatry in a vulnerable professional position, because there was no reason that medically trained practitioners should be expected to have any special expertise in treating psychologically based anxiety conditions.[37]

Freud's descendants generally abandoned his initial somatic and symptom-based theory of anxiety and used his later theory as a touchstone for their own efforts. The result was a focus on general rather than specific anxiety conditions, on interpersonal relationships and external sources of danger more than somatic processes, and on environmental as opposed to instinctual roots of anxiety. After Freud's death, the study of the psychosocial roots of anxiety took center stage.

PSYCHOANALYTIC APPROACHES TO ANXIETY AFTER FREUD

The response to Freud's work was uneven and varied widely across national contexts. His theories received the warmest reception in the United States. A number of leading academics, including Adolf Meyer at Johns Hopkins, James Jackson Putnam at Harvard, and G. Stanley Hall at Clark, advocated his theories. By the late 1920s, the popular press considered psychoanalysis to be the exemplary mode of psychological therapy. Intellectuals and bohemians, in particular, flocked to psychoanalysts to discuss "diffuse anxiety, loss of identity, inability to create, unhappiness."[38]

The experiences of military psychiatrists in World War II both enhanced the popularity of psychoanalysis and changed its thrust. Much as the First World War had altered Freud's perspective on the causation of anxiety, this war led psychiatrists to view anxiety more as the product of current environmental factors than of instinctual or early childhood experiences. They came to emphasize the universality of fear and the widespread tendency of soldiers to break down under the highly stressful conditions of modern warfare. The major cause of soldiers' neuroses was the inability of their adaptive defenses to deal with the massive amount of anxiety that war naturally produced.[39] This conceptual change would have major consequences for psychiatry in the postwar period.

The golden age of psychoanalysis in the United States began when World War II concluded. Psychoanalysts had begun to flee Europe in the early 1930s, and by the early 1940s the United States had become the center of the psychoanalytic world. Psychoanalysis dominated the clinical practice of American psychiatry and the image of psychiatry in popular culture for the next thirty years. While the leading analytic figures at the time followed Freud in maintaining that anxiety was a fundamental experience for normal as well as neurotic people, their writings fundamentally transformed opinions on the nature of this emotion.

American psychiatrist Henry Stack Sullivan (1892–1949) placed anxiety at the center of the neuroses and claimed that it determined how people shaped both their own experiences and their social relationships. Sullivan, however, thoroughly altered the accepted sources of anxiety, locating its roots in the responses of other people. Infants became and remained anxious, he believed, because of their interactions with anxious caretakers, particularly mothers. They experienced this anxiety as general tension with no apparent focus or cause. Sullivan's view shifted the focus of anxiety from individuals to the interactional networks in which they participated.

A number of the psychiatrists who emigrated from Europe created, paradoxically, a very Americanized version of psychoanalysis. Among the most prominent of these was the German-born psychiatrist Karen Horney (1895–1952), who placed anxiety at the heart of her "neo-Freudian" approach. "There is," she wrote "one essential factor common to all neuroses, and that is anxieties and the defenses built up against them. Intricate as the structure of a neurosis may be, this anxiety is the motor which sets the neurotic process going and keeps it in motion."[40] Horney's take on anxiety bore little resemblance to Freud's; it was suffused with cultural influences. In particular, she rejected the classical Freudian view that sexual forces underlay the neuroses except in "exceptional cases."[41] In the short period between Freud's development of his sexually grounded theory and 1937, when Horney's *The Neurotic Personality of Our Time* was published, cultural attitudes had changed to such an extent that sexual repression no longer had major consequences for anxiety. Likewise, even such phenomena as the Oedipus complex, which Freud had considered universal products of family processes, were instead associated with cultural conditions that generated hostility among family members.

Horney asserted that societies, not sexual or aggressive instincts, created both interpersonal conflicts and anxious personalities, which in turn led to culturally patterned responses to anxiety. Anxiety underlay neurotic character structures, which were formed in childhood through parental incapacity to provide love and nurturance. Modern life and child-rearing practices produced people who had at the core of their characters a basic anxiety, "a feeling of being small, insignificant, helpless, deserted, endangered, in a world that is out to abuse, cheat, attack, humiliate, betray, envy."[42] Horney believed that anxiousness drove the characteristic individual in the mid-twentieth century to be hostile, competitive, inferior, and emotionally isolated. Neurotics rarely confronted these feelings directly but instead projected them onto objects, situations, political events, or indistinct feelings of doom. Likewise, they employed a number of defenses against these feelings, such as the quest for affection from others, power, prestige, and material possessions. These interpersonal and intrapersonal conflicts were neither innate nor psychic but stemmed from the cultural contradictions of American life.

Horney, unlike Freud, focused particularly on the situation of women in modern societies. She found them to be in especially precarious positions because of social norms that placed numerous limitations on female achievement. Her work was especially relevant to the lives of postwar housewives, who lacked meaningful social roles as they raised their children in isolated suburban communities while their husbands worked long hours away from home.

Another popular neo-Freudian émigré from Germany, Erich Fromm (1900–1980), propelled the study of anxiety even further in a social direction. Fromm's major project was to merge Freud's insights with those of Karl Marx. His most popular work, *Escape from Freedom* (1941), treated anxiety as the central problem of modern society. In contrast to Freud, Fromm rooted modern anxieties in the particular conditions of each society rather than in biological or psychological universals. "Individual psychology," Fromm asserted, "is fundamentally social psychology."[43] For example, the Oedipal complex was not universal but a product of patriarchal societies, which placed sons in competition with their fathers because they would inherit their property after their death.

For Fromm, anxiety in the modern world was a product of unchecked capitalism, periodic structural unemployment, overpopulation, the potential for nuclear holocaust, and a host of other social ills. Lacking the security of encompassing belief systems as in earlier societies, individuals turned to totalitarian movements that protected them from the anxiety, isolation, and loneliness that freedom engendered. The key problem people faced, thought Fromm, was not their instincts or psychic repressions but understanding and overcoming a repressive society. By the 1960s, psychosocial causes had almost completely displaced hereditary and neurological explanations as the reasons for anxiety.

THE DECLINE OF PSYCHOANALYSIS

In the late 1960s the analytic view of anxiety began a steep decline that accelerated during the next decade. Psychiatry had been the institutional home of psychoanalysis, but the profession now turned sharply against the analytic view of anxiety and other disorders. One reason was the growing power of psychiatric researchers, who scorned the analysts' casual attitude toward measurement. Another was the rise to prominence of biological psychiatrists, who disparaged the analysts' focus on the unconscious, their lack of empirical rigor, and their neglect of the brain. The dominance of the analytic approach in academic departments of psychiatry faded and the number of psychiatrists identifying with psychoanalysis drastically shrunk. Once their institutional base in psychiatry crumbled, analysts fled to independent institutes, which opened themselves to all comers, not just physicians. Once psychiatry severed its ties with psychoanalysis, the prestige of psychoanalysis plummeted.

Psychoanalysis lost its social, as well as institutional, base. During the 1960s, the intellectuals and bohemians who had been among the most enthusiastic promoters of analysis in the 1920s and 1930s turned sharply against it. Far from identifying psychoanalysis with sexual liberation, rebels in the 1960s saw it as promoting conformity to repressive sexual and social roles. To them, analysis had abandoned the radical qualities that had provided much of its former allure. In addition, psychoanalysis had traditionally been identified with Jewish clients and practitioners but lost much of its appeal to Jews as they became increasingly assimilated within American society.[44]

The growing popularity of drug treatments also facilitated the fall of psychoanalysis. Although analysts often employed drugs in their practices, the use of medication was difficult to reconcile with their core theoretical tenets. One principle of psychoanalysis was that anxiety was therapeutically essential, because it motivated patients to explore their repressed thoughts and experiences. The tranquilizing drugs (discussed in Chapter 7) that became widely used in the 1950s and 1960s reduced the amount of experienced anxiety. This chemical suppression of anxiety could discourage the search for the underlying causes of neuroses. As the psychiatric profession came to organize itself around drug treatments, the influence of Freudian views of anxiety became marginalized. The analytic treatment of anxiety also became a luxury: why should people undergo long, expensive, and painstaking periods of analysis when a pill could provide them with immediate relief from their pain? Psychoanalysis couldn't match the ease, efficiency, and symptom relief of medication-based therapies.

The expense and inefficiency of psychoanalysis also led public and private funders of treatment to cast a wary eye on it. Third party payers became skeptical of the value of analytic therapies and became less willing to pay for them. As the proportion of patients using insurance and public funding to pay for treatment grew exponentially during the 1960s, economic support for psychoanalysis weakened and this therapy became limited to the small number of people who had the resources, time, and desire to pursue it.

Moreover, the movement of psychoanalysis, especially its neo-Freudian branch, toward a focus on social and environmental problems became politically toxic. Psychiatrists made significant contributions to social policy during the 1950s and 1960s through their theories of psychosocial influences on anxiety. This focus boomeranged, however, when more conservative, Republican politicians assumed power in the late 1960s. The Nixon administration directed federal funders of mental health research and policy to turn away from the study of social problems and toward the treatment of disease conditions. By the 1970s the psychosocial study of anxiety had become a politically unpalatable topic.

Freud and the theories he developed and inspired were the towering influences on the study of anxiety from the turn of the nineteenth century through the 1960s. Subsequently, the refusal of psychoanalysts to shed their insularity, to accept scientific norms, and to adapt to the biological turn in psychiatry resulted in their professional marginalization. By 1990, psychoanalysis was virtually extinct within psychiatry and just one of numerous subcultures within the mental health professions generally.

CONCLUSION

Ironically, the legacy of Freud's conception of anxiety survives to the greatest extent in what are otherwise profoundly anti-Freudian documents, the *DSMs* from 1980 to the present. The manual's various anxiety conditions and their symptoms reflect Freud's early works. By contrast, Kraepelin, generally considered the founding father of the diagnostic approach in recent *DSMs*, paid little attention to anxiety aside from its presence in other conditions, such as manic depression or schizophrenia. Although Freud's etiological claims did not influence psychiatry after *DSM-II* (indeed, Freud himself came to partly reject his early claims that all anxiety conditions had an underlying sexual etiology), current psychiatric conceptualizations of anxiety reflect his diagnostic observations far more than Kraepelin's.

Freud's other lasting contribution was to successfully expand the range of medicalization beyond conditions thought to have a physical basis. Even though neuroses resulted from interactions between individuals and their human environments and no physical pathogen needed to be present, psychiatrists came to have legitimate claims to treat conditions that had previously been viewed as purely psychogenic in nature and therefore outside the medical or neurological realm. Freud, far from his initial careful delineation of a variety of anxiety syndromes, ultimately saw anxiety as a nonspecific and ubiquitous feature of many forms of distress. Because his later theory described anxiety as present in nearly all types of normal as well as disordered conditions, most mental health outpatients could be seen as experiencing some kind of anxiety state.

Eventually, psychoanalysis became regarded as a thoroughly impractical mode of treatment. Drugs provided much more immediate and efficient relief for anxiety than lengthy periods of analysis that had uncertain results. In addition, over the course of the twentieth century the field of psychology developed a new model for treating anxiety. The emergence of behavioral theories and therapies provided a major alternative to the mentalist concepts that dominated analytic models and practice. Clinical psychologists developed techniques for the control of anxiety conditions, especially phobias, that were time-limited, easy to apply, and, at least in the short run, highly effective. Drug and behavioral treatments were more suited to the social context of anxiety that has characterized American society since the 1970s.

CHAPTER SIX

Psychology's Ascendance

AN ALTERNATIVE PARADIGM OF ANXIETY

During the first six decades of the twentieth century, psychoanalytic approaches dominated the study of anxiety within psychiatry. Outside of medicine, however, psychologists developed a behavioral view that contrasted with the biological and intrapsychic conceptions that psychiatrists had favored since the middle of the nineteenth century. The behavioral model diverged in almost every possible way from the psychodynamic approach.

For one thing, behaviorists used thoroughly positivistic methods, concentrating on observable and measurable phenomena. They scorned nonquantifiable processes and exclusively focused on objective, empirical,.and calculable measures. This starkly contrasted with the psychodynamic focus on unobservable processes such as the unconscious or the Oedipal complex. For another thing, the behaviorists took a strictly environmental approach to human action. For them, people acted in response to external rewards and punishments, not to interior biological or psychological forces. Behavior was learned through exposure to these stimuli, which differed for each individual, so all people displayed unique reactions. The behaviorists' "blank slate" conception of human nature thoroughly contrasted with psychodynamically oriented views.[1]

Finally, behavioral treatments were intensely practical. Behaviorists were less interested in explaining behavior than in transforming it; they sought behavioral change, not intellectual enlightenment. Their therapies ignored attempts to gain insight but instead sought to adjust patients to their environments. They believed that all trained practitioners could apply their techniques in uniform and standardized ways. Moreover, they claimed to change unwanted fears quickly, efficiently, and cheaply. Behaviorism thus confronted psychiatry in the therapeutic marketplace as well as in a clash of distinct worldviews. The behavioral emphasis on quantification, environmental regulation of human action, and useful results had a particular appeal to those twentieth-century Americans who felt they had no time to waste on lengthy explorations of their inner selves.

Despite vast theoretical and practical differences between behavioral and analytic approaches, one factor united them: the centrality of anxiety. In each view, anxiety was both the major reason neurotic symptoms developed and the major target of therapy. For most of the twentieth century, behaviorism was the dominant approach within psychology. Over the course of the century, the behavioral study and treatment of fear and anxiety helped propel this profession to a prominent role in American society. From John Watson's creation of a phobia in Little Albert, to the regulation of natural fears among soldiers in World War II, to the development of behavioral and cognitive treatments for anxiety disorders, psychology's ascendance was in large part due to its success in managing anxiety.

THE INSTITUTIONAL CONTEXT OF PSYCHOLOGY

The psychological study of anxiety in the United States developed in very different institutional contexts than had psychiatry. For the first half of the twentieth century, psychologists faced a major professional limitation: psychiatrists monopolized the provision of psychotherapy. In clinical settings, psychiatrists typically supervised psychologists, creating a strong source of structural tension; the former steadfastly protected their prerogatives and opposed the latter's efforts to gain clinical licensing privileges. Psychologists were generally limited to administering psychological tests and providing diagnoses. Attempting to gain some clinical independence, psychologists tended to practice in settings such as child guidance clinics, industry, and, especially, schools, locations where they did not threaten medical dominance.[2]

Psychologists played critical roles during World War I, especially in developing and applying intelligence and psychological tests. Between 1919 and 1939 the size of the profession increased tenfold, from just 300 to about 3,000 members. By 1929, the United States was home to more psychologists than the rest of the world combined; most of them were employed in applied psychology positions in education, business, courts, and clinics where they tested and evaluated children, employees, or delinquents. From its inception psychology also had a far higher percentage of female practitioners than psychiatry had, although few of them assumed leadership positions in the profession.[3]

World War II marked a decisive turning point for psychology. Until then, few psychologists had delivered psychotherapy. During the war, psychologists constructed psychological screening tests and administered them to millions of draftees. They also became involved in efforts to sustain troop morale, engage in psychological warfare, foster leadership within the military, and assign personnel to their optimal positions based on aptitude. An especially important role of wartime psychologists was to teach soldiers how to deal with their fears. Military institutions obviously placed great importance on controlling fear among the troops. While psychodynamic clinicians were also deeply concerned with fear, their intensive and lengthy therapies were highly impractical under wartime conditions. Psychologists, in contrast, had started to develop efficient and brief techniques for fear reduction.[4]

Psychologists trained combatants to endure and control their anxieties. They stressed that fear developed naturally in the face of danger, and they encouraged the open sharing of concerns with fellow soldiers, commanders, chaplains, and doctors. For example, psychologists E. G. Boring and Marjorie Van de Water wrote a best-selling book, *Psychology for the Fighting Man*, which sold nearly 400,000 copies. They emphasized how virtually all soldiers could expect to experience fearful emotions before and during combat and they pointed out the advantages of these emotions: "Fear is the body's preparation for action. The heart pounds faster, pumping blood more rapidly to the arms and legs and brain, where the oxygen is needed . . . . Subtle changes in body chemistry, automatically effected by powerful emotion, serve to protect the soldier in action in ways he would never think of, if he had to plan for himself."[5] The notion that fear was a normal response to an abnormal situation became a truism.

Psychologists in Great Britain and the United States also conducted noteworthy research about the responses of civilians and soldiers to wartime conditions. Their studies of London residents who faced repeated bombing raids demonstrated a remarkable level of suppleness among noncombatants: their fears generally subsided within an hour of these raids, and even the acute fears that a minority developed usually dissipated over a few days or weeks. Psychologists also helped develop conditioning programs for civilians, such as air-raid drills that helped desensitize them from experiencing fear in the face of aerial attacks. Their studies of combatants showed that, while soldiers nearly universally experienced a variety of anxiety symptoms, relatively few suffered psychological breakdowns. This led psychologists to emphasize the resilience of most people even in the face of extraordinarily dangerous situations.[6]

The post–World War II period was a golden age for psychology. Even if most soldiers successfully overcame their fears or recovered quickly, the number of veterans with severe and persistent psychological problems overwhelmed the capacity of the small number of psychiatrists available to treat them. In 1945 psychologists became licensed to provide unsupervised therapy, but the clinical wing of the profession was severely undersized. Immediately after the war, Congress passed the National Mental Health

Act, which included funding for preparing more clinical psychologists. Both the Veterans Administration (VA) and the National Institute of Mental Health (NIMH) began supporting graduate training programs in psychology. The roles of these psychologists expanded from mostly testing to research and therapy. By the end of the 1950s, psychologists far outnumbered psychiatrists. The size of the profession, as measured by membership in the American Psychological Association (APA), grew from 2,739 in 1940 to 30,839 in 1970, a more than tenfold increase. By 1993, membership had risen to about 75,000.[7] As increasing numbers of clinical psychologists entered private practice, conflicts with psychiatrists inevitably grew.

Behaviorists, trained in academic departments of psychology, did not rely on medical concepts or require any medical background. Behavioral studies of anxiety were based on experiments rather than clinical observations, on careful measurement and empirical validation rather than ingenious interpretations of case studies, and, most importantly, on the environmental shaping of behavior rather than on insights about the meaning of thoughts and emotions. Behaviorists' maxims—experimentation, measurement, and quantification—could hardly be more distinct from the unquantifiable, nonexperimental, and intuitive nature of psychoanalysis. Behaviorism provided psychologists with the theoretical tools to emancipate themselves from dominance by the medical profession.

BEHAVIORAL THEORY

Behavioral psychology and its views about anxiety originated in the late nineteenth century. A Leipzig psychologist, Gustav Fechner (1801–1887), wrote the first psychological textbook, which emphasized the quantitative study of what he called "psychophysics." Drawing on Fechner's work, the German psychologist Wilhelm Wundt (1832–1920) launched the first laboratory devoted to psychological studies, at the University of Leipzig in 1879. Assuming that minds had physical sensations that could be precisely measured, Wundt conducted pioneering experiments on the relationship between physiological stimuli and psychological processes.

Wundt's insistence that psychology was a rigorous experimental science continues to the present. Subjects that were not amenable to experimental methods were considered to be outside the legitimate scope of the discipline. "The laboratory," historian James Capshew states, "assumed an almost religious significance in this period as psychologists fervently asserted the claims of science in the study of human nature."[8] Psychology's allegiance to the natural sciences, combined with its focus on laboratory experiments, clearly differentiated the new discipline from its roots in philosophy, as well as from psychoanalysis.

William James (1842–1910) established the study of psychology in the United States when he became the first professor in the field, in 1876 at Harvard University. His *Principles of Psychology* (1890) set forth a scientific basis for the new profession, largely modeling it on the laws and methods of natural science. James set aside metaphysics in a quest for knowledge that was universal, objective, and, most important, useful. He was not so much concerned with the study of consciousness in itself as with how consciousness functioned in the world.[9]

James developed a new and influential theory of fear and other emotions, which came to be called the James-Lange theory, because Danish psychologist Carl Lange independently formulated the same principle. In an 1883 article, "What Is an Emotion?" James stood on its head the common notion that states of mind produced bodily expressions: "we lose our fortune, are sorry and weep; we meet a bear, are frightened and run; we are insulted by a rival, are angry and strike. The hypothesis here to be defended says that this order of sequence is incorrect. . . . We feel sorry because we cry, angry because we strike, afraid because we tremble, and not that we cry, strike or tremble because we are sorry, angry or fearful."[10] For James, physiological changes preceded and produced emotional feelings.

G. Stanley Hall (1844–1924), who became the founder of the

American Psychological Association, proposed an evolutionary view of fear. He emphasized that the most common contemporary fears were actually remnants of responses to the types of objects and situations that were common when humans evolved in pre-history but that no longer posed threats: "Night is now the safest time," Hall wrote, "serpents are no longer among our most fatal foes, and most of the animal fears do not fit the present conditions of civilised life; strangers are not usually dangerous, nor are big eyes and teeth; celestial fears fit the heavens of ancient superstition and not the heavens of modern science."[11] Psychological views of fear and anxiety, however, turned in a vastly different direction than James's and Hall's biologically based writings suggested.

In the early 1900s, John Watson (1878–1958) inaugurated the behaviorist study of anxiety in the United States. Watson went far beyond James's initial turn away from the introspective study of consciousness. He strove to banish the study of unobservable phenomena—including all inner mental states—from the legitimate realm of psychological study. For Watson, science could only concern itself with directly measurable subjects. Indeed, the study of consciousness was as irrelevant to psychology as to chemistry or physics. "Psychology as the behaviorist views it," Watson boldly stated, "is a purely objective experimental branch of natural science. Its theoretical goal is the prediction and control of behavior. Introspection forms no essential part of its methods, nor is the scientific value of its data dependent upon the readiness with which they lend themselves to interpretation in terms of consciousness."[12]

Watson did not deny that conscious processes existed but he claimed that they were not measurable and therefore were irrelevant to the scientific study of human behavior. He was impressed by the Russian neurophysiologist Ivan Pavlov's famous experiments showing how dogs could be conditioned to associate the ringing of bells with food by his repeatedly ringing a bell at their feeding times. Eventually, the dogs would salivate whenever the bells were rung, regardless of whether they were fed. The study of conditioning became foundational for Watson's behavioral psychology, which focused on how the association of past behaviors shaped future actions.

Watson was also intensely concerned with debates about the relative influence of nature and nurture in human development. Although he initially granted that a small number of emotional states, limited to fear, rage, and love, were innate, his later works denied the influence of even those instincts, focusing instead on the conditioned nature of all human behavior. Watson was an extreme environmentalist who famously declared: "Give me a dozen healthy infants, well-formed, and my own specified world to bring them up in and I'll guarantee to take any one at random and train him to become any type of specialist I might select—doctor, lawyer, artist, merchant—chief and, yes, even beggar-man and thief, regardless of his talents, penchants, tendencies, abilities, vocations, and race of his ancestors."[13] People displayed the behaviors that their environments reinforced, thought Watson; when reinforcement ceased, so did the reinforced behaviors. Abnormal behaviors did not require distinct explanations but also arose from external reinforcements, not from innate drives or preferences.

Behavioral psychology's foundational research was of fear and anxiety. Phobias, in particular, were the perfect vehicle through which psychologists could establish conditioned effects, because they were usually well-defined and easy to measure. In one of the best-known cases in psychology, Watson and his assistant (and later his wife), Rosalie Rayner, exposed an eleven-month-old boy, Albert B, to a variety of small animals, including rats, rabbits, dogs, and monkeys; the boy showed no fear in their presence. However, they discovered that loud clanging sounds from striking a hammer on a steel bar caused the boy to cry, tremble, and display labored breathing. Watson and Rayner then paired the loud noise with exposure to the previously unfeared rat. They did this seven times over a one-week period, leading Albert to whimper and cry each time. They then introduced Albert to just the rat, but not the noise. They found that he continued to display his aversive response, which they felt indicated that he had acquired a conditioned fear of rats.[14]

The experimenters claimed that Albert subsequently developed fears of not just rats but also warm, furry animals more generally and even of similar stimuli such as a Santa Claus mask and the experimenter's hair. They believed that they had demonstrated not only that fear could be a conditioned response but also that the learned reaction could be generalized to a class of related objects. "It is probable," Watson and Rayner concluded, "that many of the phobias in psychopathology are true conditioned emotional reactions either of the direct or the transferred type."[15]

The case of Little Albert initiated a long tradition of examining how environmental rewards and punishments led to conditioned fears. Despite the fact that the study used just one case, employed wholly subjective measures, and often failed to be replicated, it became one of the most-cited experiments in the history of psychology. "As much as Pavlov's dogs, Skinner's pigeons, and Milgram's obedience experiments," several psychologists proclaim, "the conditioning of Albert is the face of psychology. To many, Little Albert embodies the promise and, to some, the dangers inherent in the scientific study of behavior."[16]

Watson himself left the academy after a romantic scandal when he was only 41 years old and made no substantial contributions to behavioral theory or practice after that time. Nevertheless, his work was extraordinarily influential. The behaviorists that followed him continued to believe both that fears could be conditioned and deconditioned and that it was unnecessary to examine underlying psychological states in order to change behavior. They maintained Watson's insistence on examining only external actions. For example, one prominent behaviorist, Andrew Salter, baldly stated:

> People tell me what they think, but this does not concern me very much. I want to know what they *did*, because it is what they do that gets them into trouble, and what they *will do* that gets them out of it. To change the way a person feels and thinks about

himself, we must change the way he acts toward others; and by
constantly treating inhibition, we will be constantly getting at
the roots of his problem.[17]

Psychologists focused on the environmental circumstances that
facilitated anxiety, assuming that any stimulus could produce
fear. "Both anger and fear responses," an influential psychology
textbook stated in 1935, "are easily attached to a new or different
stimulus."[18] Even infants were said to have environmentally conditioned, rather than instinctual, responses: anxious personalities
were learned, not inherited.

Gradually, behaviorism started to expand beyond Watson's
simplistic notions. For instance, O. Hobart Mowrer (1907–1982)
studied how people developed avoidance reactions that reduced
their exposure to fearful situations. He emphasized how painful
events caused fear, which then led people to avoid similar unpleasant events in the future in order to reduce their chances of
experiencing fear. His connection of fear and avoidance had a lasting influence on later behavioral work that showed how avoiding
dreaded objects and situations could turn fears into phobias.[19]

B. F. Skinner (1904–1990), who dominated American psychology from the 1950s through the 1970s and who also became a
prominent public intellectual, was Watson's best-known descen-
dant. Skinner focused on the process of operant conditioning,
which assumed that behaviors emerged and persisted because of
their consequences. Like Watson, Skinner felt that mental activity was unobservable, and thus irrelevant for scientific purposes;
therefore, he focused on overt behaviors.[20] Unlike the Freudians,
who believed that symptoms of anxiety were merely the tip of
an underlying iceberg of anxiety, Skinner believed that symptoms
themselves were the only relevant phenomena. Maladaptive behaviors, rather than underlying conflicts, motivations, lack of insight,
or feelings were responsible for the problems people brought to
therapy. Skinner resisted theoretical formulations, instead stressing techniques of operant conditioning that treated symptoms as
being maintained by their effects. Eventually, he came to believe

that behaviorists had the power to fundamentally change all human behavior by deliberately and rationally modifying environments.[21]

Toward the end of the twentieth century, behaviorism became an intellectual dinosaur. One reason was the behaviorists' neglect of instinctual psychological processes. Behaviorists assumed that, aside from loud noises and falling, all fears were learned, so virtually any object or situation could become a source of fear. In the early 1970s, however, psychologist Martin Seligman's work showed that many fears were biologically prepared and defied the principles of learned conditioning. In fact, Seligman demonstrated that only a limited range of phenomena provoked anxiety. Echoing Darwin and Hall, he asserted that prepared fears were stimulated by objects and situations that had been genuinely fearful in ancient times, although they posed little danger at present. This inherited preparedness explained why many people remained afraid of, for example, snakes and spiders but were resilient in the face of dangerous contexts like air raids that were not evolutionarily relevant. Seligman's work forced psychologists to deal with the fact that many common fears were not learned but were instinctually preprepared.[22]

Other psychologists demonstrated that many anxious traits present in infants persisted into adulthood. For example, Jerome Kagan and colleagues found that some infants are considerably more likely than others to cry, be fearful, inhibited, attached to their mothers, and unwilling to approach unfamiliar people or objects. Their responses predicted their reactions to similar stimuli decades later. Such results indicated that anxious temperaments were largely hard-wired aspects of the brain rather than responses to environmental rewards and punishments. Behaviorist views could not encompass such findings.[23]

The behavioral emphasis on overt actions also conflicted with the cognitive turn that psychology took beginning in the late 1950s, stressing the very concepts of mind, consciousness, and meaning that Watson, Skinner, and their followers had so strenuously sought to banish from the field. Psychologists eventually recognized that learning and conditioning processes depended on the meaning of stimuli to individuals. In particular, rejecting the extreme environmentalism of Watson and Skinner, they incorporated cognitive processes as integral to the theory and treatment of anxiety. In addition to considering direct exposure and subsequent conditioning to fearful objects, psychologists started to explore how fears could be acquired vicariously through observing other people's displays of fear as well as through fear-inducing information. This shift considerably broadened the focus on conditioning and paved the way for the integration of behavioral with cognitive approaches to anxiety.

BEHAVIOR THERAPY

Psychologists were not licensed to provide therapy in the United States until the mid–1940s, so Watson's initial followers focused more on developing a new psychological science than on techniques to alter maladaptive behaviors. Nevertheless, their theories had clear relevance for behavioral change. Mary Cover Jones, a student of Watson's in the 1920s, examined how a group of institutionalized young children responded to situations such as hearing a loud noise, being left alone in a dark room, and confronting animals like snakes and rats. She then ascertained which children displayed fear responses and used a variety of methods to eliminate their fears. Jones found that two methods—direct conditioning, pairing the dreaded object with a pleasurable one, and social imitation, putting the fearful child into an environment with children who readily played with the feared stimulus—led to the extinction of the fear. Her work showed both that fears developed through the principles of learned behavior and that they could be unlearned through the same principles. Although her studies did not become widely known until the 1950s, thereafter they provided a model for behavioral therapies that emerged around that time.[24]

Especially during the 1960s, behaviorists developed a variety of therapies that used rewards and punishments to alter conditioned behaviors. They did not view anxiety as a disease but as a problem of maladaptive learning. They regarded manifestations of anxiety, such as phobias and obsessions, as being learned through the same processes as any other behavior; anxious behaviors were nonoptimal responses to environments. The behaviorists' treatments were intensely practical. They applied alternative rewards and punishments to cause people to unlearn their fears, as opposed to recovering repressed memories or prescribing pills, which would leave the reinforcers of fear untouched. Therapeutic success would result from reconditioning, not understanding or insight. It was unnecessary, they believed, to deal with any underlying causes of overt problems or with character structures. If symptoms disappeared after therapy, the problem was considered to be resolved.

The most prominent new behavioral therapies came from South African physician and psychologist Joseph Wolpe (1915–1997), using learning theory. Wolpe also trained a number of psychologists, including Stanley Rachman, Arnold Lazarus, and G. Terence Wilson, who became leading behavioral therapists in the United States and England. Like other behavioral therapists, Wolpe focused on anxiety conditions. He assumed that anxiety and other neurotic problems were conditioned; the symptoms were simply learned, maladaptive responses to previously experienced environmental stimuli. Past conditioning led organisms to become anxious in situations that would not normally create anxiety.

In contrast to Freud, who viewed symptoms as symbolic representations of unexpressed instinctual demands, Wolpe emphasized that the stimuli causing anxious responses were comparable to those that established the initial connection. He claimed that phobias began "at a particular time and in relation to a particular traumatic event." Even purported "free-floating" or "generalized" anxiety stemmed from omnipresent aspects of the environment such as light, noise, spatiality, and temporality. The pervasiveness and omnipresence of anxiety was explained by the pervasiveness and omnipresence of the stimuli that conditioned it.[25]

Wolpe believed that anxiety was at the core of neurotic behavior. Just as Watson had shown that anxious behaviors were learned, Wolpe demonstrated that they could be unlearned. He initially exposed animals to positive stimuli, such as feeding, while they experienced mild forms of anxiety. The degree of exposure to positive reinforcements would gradually increase until anxiety eventually dissipated completely. Using findings from his experiments with cats, he emphasized how neurotic reactions that developed in one context were transferred to contexts with similar features.

Based on his animal research, Wolpe developed a number of anxiety-reduction techniques using the behaviorist principle of gradual exposure to anxiety-producing stimuli. Substituting relaxation in humans for feeding in animals to inhibit anxious responses, his techniques exposed people to progressively higher doses of feared stimuli while inducing states of deep relaxation that extinguished the anxiety states.[26] Wolpe used muscle relaxation, systematic desensitization, reciprocal inhibition, "flooding" (rapid exposure to high levels of the feared stimulus), and modeling (imitating the fearless response of someone else to a stimulus). His techniques were practical, short-term, and efficient, when compared to psychoanalysis.

Behaviorists especially targeted agoraphobia—the avoidance of public places and transport because of a fear of suffering panic attacks—since it was easily measured and observed. They gradually exposed patients with intense fears of going outside and being in crowds to these situations in the presence of the therapist until they were ready to face the anxiety-provoking stimuli on their own. Likewise, those with claustrophobia might be placed in enclosed spaces, like an elevator, for gradually increasing periods of time.[27] The behaviorists reasoned that no innate instincts were involved—the people simply needed to learn how to change their behavior—so cures should be relatively quick and easy. In comparison to the years that might be necessary for successful psychoanalysis, behavioral treatments lasted for a few weeks or, at most, months. Once fears were extinguished, the problem was solved. Moreover, therapy could be completely standardized and written down, so that different therapists could use identical methods.

Hans Eysenck (1916–1997), a British psychologist who ran the Department of Psychology at the Maudsley Institute in London, also made prominent contributions to the development of behavioral theory and therapy for anxiety. He emphasized the importance of basing therapy on a strong research base: "Only rigorous and consistent theories, based on laboratory studies and generating predictions testable in the laboratory and the clinic, are worthy of detailed treatment."[28] Eysenck was particularly confrontational toward psychoanalysis. Analysis was an especially inviting target, because it not only relied on unobservable and unmeasurable concepts but also typically required years of treatment. Quick and efficient behavioral treatments thus provided a stark contrast with the lengthy periods of dynamic therapy. Sociologist Nikolas Rose summarizes Eysenck's strategy:

> The attack was to be mounted on two fronts: against the medical hegemony jealously guarded by psychiatrists and the psychotherapeutic hegemony protected by psychoanalysts. Behaviour therapy would compete with psychoanalysis at the level of its diagnostic techniques, theoretical codes, and treatment modalities. Thus psychoanalysis was to be discredited at all three levels.[29]

Most notably, in 1952 Eysenck published a polemic against psychoanalysis. He evaluated the impact of psychoanalysis, concluding that it was no more effective than not entering therapy at all. In a later article, Eysenck maintained his earlier conclusion but added data he thought indicated that therapy based on learning models was more effective than psychotherapy. For Eysenck, behavior therapy was a superior method of treatment while psychotherapy had virtually no clinical relevance.[30]

Although Eysenck's data, especially his estimate of what proportion of patients would recover in the absence of therapy, were questionable, his studies put analysts on the defensive. Behaviorists asserted that phobias, compulsions, and obsessions in particular could be cured effectively, quickly, and cheaply. They contrasted their own reportedly high rate of therapeutic success with what they characterized as the dismal results of analysis. Wolpe, for example, claimed a 90 percent success rate from the use of his relaxation-desensitization approach.[31] The Achilles heel of psychodynamic treatments became the issue of their questionable therapeutic effectiveness.

Despite protests from psychoanalytical clinicians that symptoms behaviorists thought they had extinguished would reappear in different guises, that behavioral techniques were mechanized and dehumanized, and that claims for their effectiveness were inflated,[32] behavioral therapies successfully became institutionalized. Behaviorism became the dominant view in many prominent universities and research institutes, especially in the English-speaking countries of the United States, England, and South Africa. During the 1960s, several professional psychological societies formed, including the Association for Advancement of Behavior Therapy and the Behavior Therapy and Research Society.[33] In the 1960s and 1970s many journals devoted to disseminating behavioral therapy's results were established, including *Behaviour Research and Therapy, Behavior Therapy, Journal of Behavior Therapy and Experimental Psychology, Journal of the Experimental Analysis of Behavior,* and the *Journal of Applied Behavior Analysis.* Regular conferences, symposia, and workshops contributed to a thoroughgoing professionalization of behavioral therapy during these decades. Gradually, behaviorism overcame Watson's scientific dogmatism to become more eclectic and open to a diversity of therapeutic approaches, especially those that focused on changing faulty cognitive processes.

COGNITIVE THERAPY

While behavioral methods were major competitors to psychodynamic therapies in the treatment of anxiety during the 1950s and 1960s, since that time cognitive approaches have emerged as the major form of therapy that does not rely on medication. By the 1980s cognitive approaches had supplanted pure behaviorism as the dominant theoretical view within psychology. In contrast to the behaviorists, cognitivists embraced the study of the mind. In the late 1950s, as a response to criticisms that behavioral therapy ignored the critical aspect of cognition, psychologists had begun to develop cognitive behavioral therapy (CBT). Their core assumption was that false beliefs and mistaken patterns of interpretation of events, rather than environmental reinforcers, underlay fear and anxiety disorders. Cognitive therapies strive to change internal perceptions, exactly the sort of phenomena that behaviorists studiously avoided.

Cognitive treatments are grounded in rationalistic philosophy.[34] Their basic postulate is that normal people think rationally about the world, so therapy should focus on training the mentally ill to think accurately and logically about the world. Reason, according to this view, can tame emotions and instincts. There is no need to deal with the unconscious, innate biological drives, or memories of past traumas. Nor is social reality problematic, because individual perceptions, rather than the social world, must be changed. At least implicitly, CBT actively strives to get patients to conform to the social status quo and to think in the same ways as most other people do.

Psychologist Albert Ellis (1913–2007) used the principles of behaviorism but applied them to the quite different sphere of thinking. For Ellis, maladaptive behaviors resulted from incorrect perceptions of reality. He developed rational-emotive therapy, which is based on the principle that irrational patterns of thought produce anxiety (and other mental) disorders. The major task of this therapy is to challenge and change the irrational beliefs that are presumably the basis for unreasonable fears and anxiety.[35] Aaron Beck (1921–) is another major figure in the development of CBT. Although Beck trained in psychiatry, his work has had little influence in that field. While Beck is best known for his work on depression, he also has written extensively on anxiety.

In contrast to psychodynamic approaches, CBT focuses on present problems, not past traumas. These problems are considered to be fully accessible to conscious thought, not hidden within the unconscious and therefore difficult to recover. Thinking controls behavior, reason cognitivists, so if therapists can change the way their patients think, they can resolve their problems. Distorted internal belief systems that provide incorrect views of the external world produce psychopathology. Beck wrote about a typical anxious patient, saying that his fears were based "on his overinterpretation of 'danger' stimuli, his distortions of the incoming stimuli, and his arbitrary inferences and overgeneralizations."[36] Anxiety thus develops when people interpret the world through erroneous patterns of thinking that make it seem more threatening than it actually is and as a result make dysfunctional responses to perceived threats. Effective therapies alter these distorted thought patterns and teach patients how to think rationally about their actual situations. Changing the cognitive processes responsible for anxiety should eradicate the anxious condition itself.

CBT is a far shorter process than psychoanalysis, usually lasting between several weeks and rarely exceeding several months. Moreover, it can be reduced to specific, structured, and standardized procedures that are embedded in manuals. Any trained practitioner can thus be an effective therapist, because the method more than the practitioner leads to success. Theoretically, the personal qualities of the therapist are irrelevant: the therapist is simply a highly trained expert who applies a particular set of techniques. The development of CBT, rooted firmly in evidence-based practice, provided clinical psychologists with a powerful method for achieving professional success and prestige. Cognitive theories and therapies constitute the most potent current alternative to biological approaches to anxiety.

CONCLUSION

Behavioral and cognitive approaches to anxiety and its treatment became tremendously successful in the English-speaking world during the latter half of the twentieth century, helping to establish the scientific credentials of psychology as a discipline and to provide it with credible alternatives to psychodynamic therapies. Psychologists built an efficient set of treatments, centered on the control of anxiety, based on behavioral theory. Their therapies were far more practical than those of the psychoanalysts. For much of the twentieth century, the analytic and behavioral paradigms of anxiety uneasily coexisted. If the former gained more cultural glory, the latter possessed more impressive therapeutic credentials. By the end of the century, however, a new biological paradigm, restating many of the themes of its nineteenth-century predecessors, steamrollered both alternatives to dominate the study and treatment of anxiety.

A major key to the success of behavioral and cognitive treatments lay in their greater practicality compared to the prolonged process of analysis. Advocates justified CBT and other behavioral treatments by their efficiency and cost-effectiveness. Yet, the widespread use of drug therapies for anxiety eventually hoisted the behaviorists on their own petard. Ingesting a benzodiazepine or, later, an SSRI was far easier and less time consuming than visiting a behavioral or cognitive therapist. By the end of the century, numerous studies had concluded that, while CBT in particular was an effective response to anxiety, it was no more effective than medication treatments.[37] The managed care companies that controlled reimbursement for medical costs believed that drugs were considerably cheaper than other therapies. Over the course of the 1990s and the first decade of the twenty-first century, rates of CBT began to decline while the use of medications soared.[38]

The biological turn in psychiatry at the end of the century also called into question the roles of learned behavioral conditioning and faulty ways of thinking as generators of anxiety. Nurture lost ground to nature when the dominant scientific lens turned toward the study of genes, neurotransmitters, and brain circuitry. While behaviorism integrated cognitive factors into its model of anxiety, it was less compatible with the movement back toward looking at anxiety through biological lenses. At the end of the twentieth century the revival of biological approaches to the study of anxiety coupled with the growing power of pharmaceutical companies trumped the success of psychological treatments for anxiety. As was the case in the nineteenth century, anxiety again became thought of as grounded in somatic processes, not behavioral re-inforcement or cognition. Before a biological model could be re-implemented, however, a revolution in the conceptualization of anxiety had to occur. The publication of the *DSM-III* in 1980 paved the way for a new model of anxiety that overcame not just psychoanalysis but, ultimately, behaviorism as well.

CHAPTER SEVEN

The Age of Anxiety

ANXIETY BECOMES UBIQUITOUS

Through much of the twentieth century, anxiety was not only a central condition in medical and psychiatric practice but also a major aspect of Western culture. The first sentence of Rollo May's popular book, *The Meaning of Anxiety* (1950), was illustrative: "Every alert citizen of our society realizes, on the basis of his own experience as well as his observation of his fellow-men, that anxiety is a pervasive and profound phenomenon in the twentieth century."[1] The prominence of anxiety was largely due to Freud's immense popularity among intellectuals and the media. Anxiety was also a central topic in the works of popular existential philosophers and liberal theologians, including Martin Heidegger, Jean-Paul Sartre, Martin Buber, and Reinhold Niebuhr. Existentialism, in particular, placed anxiety at the heart of the human predicament.[2]

For most of the century, anxiety was not usually formulated as a disease. Analytic and philosophical views typically viewed anxiety as an unfortunate but inescapable component of existence. May's book, for example, used an epigraph from Freud: "There is no question that the problem of anxiety is a nodal point at which the most various and important questions converge, a riddle whose solution would be bound to throw a flood of light on our whole mental existence."[3] During the 1950s and 1960s, anxiety was such a common aspect of American culture that the era itself became known as the Age of Anxiety.[4]

Another common nonmedical view was that anxiety was a natural response to social circumstances that required collective, political solutions. The ubiquity of anxiety in the post–World War II era was attributed to the horrors of the war, the development of nuclear weapons, and the potentially catastrophic tensions of the cold war between the United States and the Soviet Union. Not just worry about global problems but also the results of everyday stressors—whether problems with spouses and children, parents and bosses, or overwork and daily hassles—were called anxiety. Anxiety became as emblematic a symbol of the collective emotional state of mid-twentieth-century Americans as neurasthenia or neurosis had been in the late nineteenth and early twentieth centuries and as depression would become in later decades. The feminist author Betty Friedan captured the capacious aspect of this condition when she labeled the characteristic psychic dilemma of middle-class women as "the problem that has no name," which featured "mild, undiagnosable symptoms . . . malaise, nervousness, and fatigue."[5] George Beard's neurasthenics would have been familiar with this ailment. "Anxiety" was, in essence, a synonym for "distress."

The labeling of distress as "anxiety," along with the emergence of a therapy-oriented culture in the post–World War II period, led to a huge demand for relief from anxious conditions. Despite its strong influence on American culture, psychoanalysis was actually a tiny profession. In 1957 only about 1,000 analysts practiced in the United States.[6] Such a minute group could hardly deal with more than a small fraction of the treatment that was being sought by anxious people. With the emergence of anti-anxiety medications during the mid–1950s, people started turning to general physicians far more than to psychiatrists for relief. Anxiety was the target of the first generation of blockbuster medications, the minor tranquilizers. These drugs became an object of consumerism, promoted in the mainstream and tabloid press, mass circulation magazines, and the new medium of television. Like automobiles, appliances, and suburban homes, the tranquilizing drugs became part of the fabric of the postwar American way of life.[7]

THE RISE OF THE TRANQUILIZERS

For centuries, people had sought relief for anxiousness through a variety of substances, including opiates, alcohol, and bromides, all of which were freely available. The twentieth century, however, ushered in a new, far more restrictive, approach to drug sales, especially in the United States. The Pure Food and Drug Act of 1906 made the over-the-counter sale of "dangerous" drugs, including opium and morphine, more difficult. Eight years later the Harrison Act criminalized the prescription of narcotics for addicts.

During the first part of the century, the pharmaceutical industry did not actively promote prescription medicines. Instead, they sold raw materials to neighborhood pharmacists, who assembled the final products and retailed them to consumers. The barbiturates were popular but were associated with a variety of negative side effects, a high potential for dependence, and were often used in suicide attempts. In 1951 the Durham-Humphrey amendments to the 1938 Food, Drug, and Cosmetics Act gave the Food and Drug Administration (FDA) the power to decree what drugs could be purchased solely through a prescription. This led to a sharp split between prescription drugs and drugs that could be purchased "over the counter." In 1962 the FDA required that drugs come to market only after undergoing randomized clinical trials that proved their effectiveness in treating recognized diseases. Although the agency did not strictly enforce the 1962 rules until the following decade, their existence influenced the recategorizing of anxiety from general distress to specific disease states. More importantly, treatments for anxiety became highly valued targets for the pharmaceutical industry, which was poised to create mass markets for its products.

Pharmaceutical companies, not university or government scientists, developed the first mass-marketed drugs for anxiety. During the 1950s, the drug industry began to focus on developing lucrative products for broad consumption. The first minor tranquilizer, meprobamate (initially marketed under the trade name Miltown by its manufacturer, Wallace Laboratories), was an explicitly anti-anxiety medication that combined tension-reduction and muscle-relaxing properties. It was intended to treat problems of anxiety among the many people who sought help for their anxiety from a general physician. Renowned psychopharmacologist Nathan Kline convinced Frank Berger, the chemist who developed Miltown, to call it a "tranquilizer." "What the world really needs," Kline urged, "is a tranquilizer. The world needs tranquility. Why don't you call this a tranquilizer? You will sell ten times more."[8]

Berger viewed tranquilizers as remedies for a biologically based condition. "Meprobamate," Berger explained, "has a selective action on those specific areas of the brain that represent the biological substrate of anxiety." He made great claims for such substances: "Drugs can . . . eliminate obstructions and blockages that impede the proper use of the brain. Tranquilizers, by attenuating the disruptive influence of anxiety on the mind, open the way to a better and more coordinated use of the existing gifts. By doing this, they are adding to happiness, human achievement, and the dignity of man."[9] For many, including its discoverer, Miltown was a drug that enhanced happiness as well as corrected a biological defect.

From their origin, the tranquilizers were linked to public relations, marketing, and advertising. Carter Products (the parent company of Wallace Laboratories) introduced Miltown to the public and to physicians in 1955 through large promotional and publicity campaigns. Although drug companies were forbidden to advertise their products directly to the lay public at the time, they found many indirect ways to reach consumers. One strategy used large sales forces that provided samples of their products to general physicians. Another used massive advertising campaigns in medical journals along with physician education programs to increase awareness of the new drugs. Drug companies also directly reached the general public by employing public relations and marketing companies to plant stories in the mass media that promoted the tranquilizers as means to control stress and become happy. The popular attitude toward Miltown in the 1950s and '60s was overwhelmingly favorable; the media was awash with claims that the drug was a miracle cure for the tensions and stresses of everyday life.[10]

Miltown was promoted far more for its ability to relieve normal problems of stress than as a remedy for specific psychiatric conditions. Advertisements touted an extremely broad array of indications for its use, including "tense, nervous" patients, "insomniacs," "anxious depression," "alcoholics," and "problem children."[11] It was, in short, the ideal treatment for the ubiquitous "stress," "nerves," or "tension" that accompanied everyday life. Harking back to Beard, this usage transformed the analytic view of anxiety as an unconscious inner force into a condition that stemmed from the general environment. One ad even encouraged physicians to prescribe Miltown so that pregnancy could "be made a happier experience." Another featured the "battered parent syndrome" suffered by a housewife for whom child care led her to be "physically and emotionally overworked, overwrought and—by the time you see her—probably overwhelmed."[12]

Miltown was an instant success. By the end of its first year on the market, fully 5 percent of Americans had taken the drug and its sales exceeded $2 million. Miltown's popularity stemmed from its easy availability, affordability, and rapid effect. Once received, a prescription could be refilled indefinitely, enhancing the ease of using it. Initial medical reports indicated remarkable effectiveness: from 65 to 90 percent of patients suffering from anxiety, fear, and tension underwent dramatic improvement or complete cure. Even most psychoanalysts, who were not generally proponents of medication treatments, prescribed tranquilizers for many patients during the 1950s and 1960s.[13]

Patients usually turned to general practitioners rather than psychiatrists for chemical relief from anxiety; GPs wrote about 80 percent of the prescriptions for tranquilizers. By 1960 about three-quarters of all physicians were prescribing Miltown or one of its equivalents. Most used the term "anxiety" to describe diffuse and nonspecific complaints. "An abundance of tensions, fears, worries and anxieties confront mankind today," prominent anxiety researcher Karl Rickels noted, "and, in fact, anxiety is seen in the majority of patients visiting the physician's office."[14] Just one-third of prescriptions were associated with a diagnosis of a specific mental illness. "Only about 30 percent of use," an article in the *Journal of the American Medical Association* (*JAMA*) observed, "is in identified mental disorders and the remainder covers the rest of medicine."[15] In contrast, about 60 percent of consultations involved problems of daily living. Patients and doctors embraced tranquilizers with enthusiasm. The growing middle class, in particular, eagerly sought the new medications. Women, especially the middle-aged and elderly, were more than twice as likely as men to consume these drugs.[16]

Miltown's great success led pharmaceutical manufacturers to search for even more lucrative successors. The benzodiazepines Librium and Valium became the next blockbuster drugs. They were considerably more potent than Miltown yet were less sedating. First Librium, and then Valium, became the single most prescribed drug in the Western world. From the 45.1 million prescriptions written for the minor tranquilizers in 1964, use soared to 103.2 million at their peak in 1975. At that point, 20 percent of all women and 8 percent of all men in the United States reported having used a tranquilizer at some point in the previous year, and about a third of this group reported being regular users.[17] Librium and Valium were, according to historian David Herzberg, "the quintessential women's drugs of the 1960s and 1970s, respectively."[18]

The promotions for the benzodiazepines echoed the earlier themes used for Miltown. Valium and Librium were hailed for their ability to ease a huge range of conditions involving nervous tension. Advertisements in medical journals touted their ability to relieve anxiety associated with the stress of problems with spouses and children, overwork, career failures, social role transitions, housekeeping, and even traffic jams. Other promotions emphasized their use for such general situations as the "relief of tension and anxiety alone or whenever somatic complaints are concomitants of emotional factors." For example, an ad campaign for Librium emphasized its use for treatment of "mild to moderate anxiety and tension, tension headache . . . and whenever anxiety and tension are concomitants of gastrointestinal, cardiovascular, gynecologic or dermatologic disorder." Another ad for Librium promoted it for the treatment of normal college life, which it portrayed thusly:

> A whole new world of anxiety. The new college student may be afflicted by a sense of lost identity in a strong environment. Today's changing morality and the possible consequences of her "new freedom" may provoke acute feelings of insecurity. She may be excessively concerned over competition, both male and female, for top grades. Unrealistic parental expectations may further increase emotional tension.

An ad for Valium explicitly advocated its use for people without any psychiatric symptoms, for instance, an overworked doctoral candidate who was finishing a thesis and had indigestion and gastrointestinal problems: "For this kind of patient—with no demonstrable pathology—consider the usefulness of Valium."[19]

The result was to thoroughly blur the borders between psychiatric illness and normal anxiety. In 1975 a review of tranquilizer use concluded: "What illnesses are being treated? Most of what primary care physicians see, they label 'anxiety.'"[20] Tranquilizers had become firmly embedded within American culture not so much as treatments for anxiety disorders as to relieve the stresses of ordinary daily life. Moreover, psychopharmacology was elevated to an extraordinary pedestal: "When medical historians name the chief advances in the second half of the twentieth century, organ transplantation and nucleic acid chemistry will rank high, but it may be that the increased understanding of pharmacological events underlying human physiology, disease, and behavior will be given an even higher place."[21] The extensive use of tranquilizers reconfigured the response to psychological problems away from talk and toward drug therapies. "Valium," historian Andrea Tone notes, "rapidly became a staple in medicine cabinets, as common as toothpaste, brushes, and razors."[22]

THE REACTION AGAINST THE TRANQUILIZERS

Ever since the introduction of the minor tranquilizers, considerable debate had occurred over their efficacy and abuse. Concerns about their mass use for relieving ordinary distress surfaced as early as 1956. A report from the Subcommittee on Tranquillizing Drugs of the New York Academy of Medicine cautioned: "Anxiety and tension seem to abound in our modern culture and the current trend is to escape the unpleasantness of its impact. But when has life ever been exempt from stress? In the long run is it desirable that a population be ever free from tension? Should there be a pill for every mood or occasion?"[23] By 1958, Surgeon General L. E. Burney cautioned that "problems of daily living" cannot be solved "with a pill."[24] In the same year, Congress convened a hearing titled "False and Misleading Advertising of Prescription Tranquilizing Drugs," the first in a series of investigations over the misuse of these drugs as treatments for normal anxiety rather than for psychiatric disorders.

While some psychopharmacologists believed that the tranquilizers were of limited efficacy, prone to abuse, and vastly overprescribed, most felt they were effective, safe, and underutilized. For them, the opposition was based on misguided, morally based aversions to drug use. Psychiatrist Gerald Klerman famously decried what he called "pharmacological Calvinism," which held that "if a drug makes you feel good it must be bad. Indeed, if a drug makes you feel good, it's either somehow morally wrong, or you're going to pay for it with dependence, liver damage, chromosomal change, or some other form of secular theological retribution."[25] Opposition notwithstanding, the tranquilizers were wildly popular from their introduction in the mid–1950s through the mid–1970s. At that point they suffered a sharp reversal in fortune.

By the early 1970s the FDA had begun to enforce its regulation that drugs must target recognized disease conditions. It assumed that mental illnesses, like most physical ones, were discrete and could be easily separated from ordinary distress. FDA Commissioner Charles Edwards sent a letter to the sponsors of all new psychotropic drug applications that stated: "These drugs are approved for the treatment of clinically significant anxiety, depression, and/or other mental conditions, but, of late, advertisements have also promoted their use in the treatment of symptoms arising from the stresses of everyday living, which cannot properly be defined as pathological."[26] Increasing regulatory pressure led drug companies to stop touting their products as remedies for the general stress of living.

A sharp reaction against the tranquilizing drugs developed in the general culture as well. Feminist groups in particular became vocal opponents of the drugs' use because of the potential for addiction to them and a perception that they were implicated in the repression of women's activities. Leaders of this movement, such as Betty Friedan, singled out tranquilizers as being means of social control that kept housewives in their place. Advocacy groups in both the United States and Europe began to mount legal challenges against the benzodiazepines. These lawsuits, which claimed that these drugs were highly addictive, drew considerable media attention. At the same time, feminist and patient advocacy groups launched public relations campaigns featuring messages about the dangers of benzodiazepine use. News stories reported tranquilizer addiction among both ordinary women and celebrities, including First Lady Betty Ford and television producer Barbara Gordon. Several highly publicized congressional hearings reinforced the impression of widespread addiction, dependence, and suicidal potential of these drugs.[27]

The stigma being placed on the tranquilizers resulted more from political and social pressures than expert opinion. Most researchers felt that the benzodiazepines were safe and effective, were rarely addictive unless abused, and were difficult to use in suicide attempts. A typical summary in one *JAMA* article stated: "In treating minor emotional states with mixtures of anxiety and

depression, the benzodiazepines are as effective as any other seda-
tive drug, they have no overdose potential; tolerance, abuse, and
abstinence are very rare; and they have remarkably few side effects."[28] Prominent psychiatrist Frank Ayd claimed that the insignificant number of benzodiazepine abusers were "psychopathic
personalities with histories of abuse of alcohol or drugs."[29]

Despite the assertions of much of the medical community,
in 1975 the FDA classified Librium and Valium as Schedule IV
drugs, which limited the number of prescription refills, required
strict reporting, mandated insertion of a statement that the drugs
could be abused, provided strong criminal sanctions for illegal
sales, and ordered the inclusion of an information packet that
stated: "Anxiety or tension associated with the stress of everyday
life usually does not require treatment with an anxiolytic (anti-
anxiety) drug." A sharp plunge in sales resulted: prescriptions for
tranquilizers in the United States plummeted from 103.2 million
in 1975 to 71.4 million in 1980. The marketers' dream drugs had
become commercial nightmares.[30]

DSM-I AND DSM-II

During the era when the tranquilizers became consumer products
that helped people cope with their distress, anxiety remained a
central aspect of psychiatric diagnosis. In particular, it was at the
core of the first classificatory manual of the American Psychiatric
Association, the *Diagnostic and Statistical Manual of Mental Disorders* (1952). The APA developed the *DSM-I* as a response to the
deficiencies of the extant *Statistical Manual for the Classification
of Mental Disorders*, which was primarily targeted for use among
inpatient populations. All except one of the *Statistical Manual*'s
twenty-two principal groups concerned psychotic and other very
severe conditions that were found mostly among residents of mental institutions. The one other category, psychoneurotic conditions, dealt with all other states of mental disorder.

The experiences of military psychiatrists during World War II
revealed that the *Statistical Manual*'s categories were not appropriate for most of the psychic problems that soldiers developed. Their

conditions seemed to better fit on a continuum that ranged from normal on one end to very abnormal on the other, rather than on one side or the other of a sharp differentiation between health and disease. As well, the fact that so many previously normal soldiers developed mental disorders in wartime seemed to point to the causative role of the environment. In addition, psychiatric practice was increasingly moving away from inpatient institutions and into the community. The existing manual, however, gave short shrift to the types of problems that an increasingly office-based specialty was treating.[31]

The *DSM-I* reflected the institutional dominance of psychoanalysts, who commanded the most prestigious positions in academic departments of psychiatry, professional organizations, and outpatient practices. It combined Freud's later views of neuroses, which downplayed the role of manifest symptoms and stressed the importance of underlying psychic mechanisms, with the psychosocial views of American psychiatrist Adolf Meyer, which emphasized how people's life experiences led them to develop neuroses.

The concept of "neurosis," a catch-all term for many nonpsychotic psychic disturbances, was at the heart of the new manual; and at the heart of the neuroses was anxiety. The very first sentence of the Psychoneurotic Disorders category stated: "The chief characteristic of these disorders is 'anxiety' which may be directly felt and expressed or which may be unconsciously and automatically controlled by the utilization of various psychological defense mechanisms (depression, conversion, displacement, etc.)."[32] Some of the psychoneuroses were explicitly described as anxious conditions, including anxiety neuroses, phobias, and obsessive-compulsive reactions. Others, such as depressive, dissociative, and conversion reactions, were viewed as defensive adaptions to unconscious anxiety.

The next revision of the manual, the *DSM-II* (1968), maintained the centrality of anxiety. Its definition of the psychoneuroses stated: "Anxiety is the chief characteristic of the neuroses." The *DSM-II* downplayed the importance of overt symptoms, stating,

"[Anxiety] may be felt and expressed directly, or it may be controlled unconsciously and automatically by conversion, displacement and various other psychological mechanisms."[33] Anxiety was
thus a capacious condition that underlay all of the neuroses—not
just the anxiety neuroses but also hysterical, depressive, neurasthenic, and hypochondriacal neuroses. The manual, however,
provided only very short and cursory definitions of each of these
conditions.

The amorphous definitions of anxiety conditions in *DSM-I*
and *DSM-II* encouraged blurry boundaries between normal and
pathological anxiety in epidemiology as well as in clinical practice.
During the 1950s and 1960s epidemiological research reflected
the shift of psychiatric attention from serious disorders to milder
cases. These studies rarely looked at specific mental disorders but
relied on broad and continuous symptom scales that emphasized
anxiety. For example, the best-known measure of symptoms, the
Langner scale used in the Midtown Manhattan Study, contained
twenty-two items. At least twelve of these were symptoms of anxiety—"restlessness, nervousness, worries getting you down, being
the worrying type, feeling hot all over, heart beating hard, shortness of breath, fainting spells, acid stomach, cold sweats, fullness
in the head, and trembling in the hands."[34]

As study data on people reporting symptoms of anxiety accu-
mulated, the division between mental normality and abnormality
became increasingly blurry. Symptom scales were viewed as broad
measures of distress that indicated conditions ranging continuously from mild to moderate to severe, depending on the number
of symptoms respondents reported. They revealed that a majority
of the population reported symptoms that indicated some degree
of mental illness, supporting prominent psychiatrist Karl Menninger's assertion, "It is now accepted that most people have some
degree of mental illness at some time."[35] In line with the prevailing
theories of the time, such huge rates of mental illness were expect-
able products of such psychosocial stressors as poverty, low social
class, poor education, and community disorganization.

The *DSM-I* and the *DSM-II* provided suitable classificatory

frameworks in the decades following World War II when clinicians paid little attention to diagnostic manuals and their practice
focused on general mechanisms rather than precise diagnoses. In
addition, most dynamic therapies did not vary across different
anxiety conditions but were applied across diagnostic categories.
Moreover, because most financial transactions in psychotherapeu-
tic relationships were directly between patient and therapist, no
third party required discrete diagnoses in order to compensate
providers. Consequently, during this era, specific diagnostic categories were not needed for intellectual, professional, or economic
purposes and so diagnosis was not a primary concern.

THE DSM-III

During the heyday of psychodynamic dominance, psychiatry had
no reasons to develop elaborate classifications. Other medical specialties, however, emphasized the importance of diagnostic specificity. The absence of a systematic and reliable nosology reinforced
the marginalization of psychiatry in the medical field. By the late
1960s and early 1970s, the lack of a satisfactory diagnostic system
was becoming a noose around the profession's neck. Many prominent critics claimed that misdiagnosis of patients was rampant.
Others, such as renegade psychiatrist Thomas Szasz, went further,
claiming that the field couldn't even define its most central concept, mental illness.[36]

As noted, the FDA was becoming less tolerant of the prescription of anti-anxiety drugs to deal with the stresses of everyday
life rather than for specific mental illnesses. In addition, as third
party reimbursement of treatment grew, private and public insurers increasingly demanded that the conditions they were paying to
treat be genuine illnesses and not normal worries and concerns. It
was becoming clear that psychiatry needed a more easily measurable and reliable system of diagnosis to preserve its legitimacy as
a medical specialty.

At the same time, a movement against psychoanalysis was developing within psychiatry. At its heart were empirical-minded
psychiatrists led by Robert Spitzer and his allies at Columbia and

Washington Universities who opposed the murky *DSM-II* diagnoses that centered on the concept of neurosis. The *DSM-I* and *DSM-II* focused on the underlying meanings and causes of various symptoms, which they assumed largely had to do with anxiety-related processes. These manuals treated any particular condition, such as panic, obsessive-compulsive disorder, or phobia, as a surface manifestation of more basic neurotic mechanisms. They also attributed conditions like psychosomatic illnesses and depression that were marked by symptoms such as headaches, ulcers, and skin rashes with no apparent organic basis to anxiety. Because anxiety underlay such a wide variety of psychoneurotic conditions, an attack on the concept of neurosis also challenged the psychoanalytical view that anxiety was the fundamental process behind neurosis itself.

Another key rationale for creating the *DSM-III* was the meager level of understanding about the causes of psychiatric conditions. Therefore, the manual's developers assumed that only manifest symptoms, not theoretical assumptions, should be used to differentiate various conditions. In each of its definitions, the new manual used explicit, symptom-based criteria that were not tied to any particular theory of disorder. The removal of etiological inferences from the diagnostic criteria helped make diagnoses useful to therapists holding many diverse theoretical allegiances, increasing the *DSM*'s potential audience.

The empiricists, relying on several sources of data, launched an ultimately successful attempt to undermine the *DSM-II* and replace it with a new and more useful manual. One of their sources was the research of a group of psychiatrists led by Samuel Guze and Eli Robins at Washington University of St. Louis, a medically oriented outpost of empirical psychiatry during the era when psychodynamic theory and practice dominated the field. These psychiatrists developed what came to be known as the Feighner criteria, named after the psychiatry department's chief resident, who had recorded the observations of the group. Among the fourteen psychiatric illnesses included in the Feighner criteria were three anxiety conditions: anxiety neurosis, obsessive-compulsive neurosis, and phobic neurosis. Symptomatic indicators, namely anxiety attacks, persistent ideas and compulsions, and recurring fears, respectively, defined each of these conditions.[37]

The Feighner criteria were a highly significant development, because they provided a specific, almost diametrically opposed alternative to the *DSM-II*'s psychodynamic conceptions of anxiety. They were much more easily measurable than the criteria defining psychoneuroses in the *DSM-II* and did not assume any particular causal structure for anxiety symptoms. They also separated the various anxiety conditions from depression and hysteria, without giving causal primacy to anxiety. Moreover, they basically abandoned the psychodynamic tradition's view of generalized anxiety as the core neurotic condition, instead linking it to panic attacks rather than diffuse anxiety. This further undercut the centrality of anxiety in the psychoneuroses more generally. Although the Feighner criteria were developed for research, rather than clinical, purposes they became a basic resource for Spitzer and his collaborators as they developed the diagnostically oriented *DSM-III*.

Psychiatrist and pharmacologist Donald Klein's work was another major stimulus behind the *DSM-III*'s eventual delineation of the anxiety disorders. In 1959, Klein observed that the drug imipramine, which was believed to be a stimulant, in fact initially led to sedation and then to enhanced appetite among depressed patients. When he administered imipramine to anxious patients, he found that it prevented panic attacks from recurring but had no impact on chronically anxious patients. This indicated to Klein that the drug differently affected two types of what had been thought to be the same underlying condition. These selective consequences seemed especially anomalous because the drug worked on the more severe, but not the less severe, variant of the condition. In a chapter titled "Anxiety Reconceptualized," Klein concluded:

> Perhaps the most important thing learned from this experience was the power of the experimental technique of pharmacological dissection whereby one can pierce through the fascinating, con-

fusing web of symptoms and dysfunctions to tease out the major participant variables by attending to specific drug effects. This led to the useful methodological conclusion that it makes sense to be an experimental splitter rather than a naturalistic lumper.[38]

Klein's work opened a pathway to dissecting a variety of anxiety conditions on the basis of their differential responses to medication.

Klein used his experience with imipramine to develop a more sweeping critique of the analytic and psychosocial models that underlay the *DSM-II*'s approach to anxiety. He claimed that those two views shared the deficiency of treating all anxiety states as being grounded in traumatic experiences. In contrast, Klein demonstrated the power of experimental pharmacology to distinguish panic attacks from chronic anxiety and suggested the potential of drug response to differentiate other forms of anxiety, including phobias and generalized anxiety. In another demonstration of Klein's theory, the common drug responses of panic disorder and agoraphobia showed that they should be treated similarly with one another but distinctly from other anxiety states.

Klein's research heralded a changing of the guard from psychoanalytic to psychopharmacological approaches to defining the nature of anxiety. It also represented a turning back to the conditions that Freud had delineated early in his career but then neglected in his later writings—panic, generalized anxiety, phobias, and obsessions. Klein's delineation of specific anxiety conditions on the basis of drug treatment response proved to be enormously influential in moving conceptions of anxiety from general and dimensional toward specific and categorical models.[39]

Psychoanalysts reacted with fury to the attack on the concept of neuroses in the formulation of the *DSM-III*. They realized that the new manual was not simply intended to sharpen the diagnostic system but was a frontal assault on the foundational assumption of psychoanalysis itself—that intrapsychic conflict related to anxiety produced most psychoneurotic conditions. "The DSM gets rid of the castles of Neurosis," one analyst memorably complained, "and replaces it with a diagnostic Levittown."[40] Oddly, "dysthymia," the term the *DSM-III* used to capture the rejected class of neurosis, was a mood, not an anxiety, disorder. Moreover, the manual's requirement that dysthymic symptoms have been present for two years meant that few persons would satisfy the criteria for the diagnosis. The analysts' protests fell on deaf ears in the *DSM-III* task force. The result was a destruction of the analytic conception of the psychoneurosis in general and of anxiety in particular.

ANXIETY IN THE DSM-III

The *DSM-III* revolutionized conceptions of anxiety. Its underlying philosophy was that psychiatry needed to become a data-driven medical specialty in contrast to the "dogma and theory" that guided psychoanalysis.[41] The new manual's core mission was to develop presumably more *reliable* criteria with which to distinguish one condition from another. This would allow different psychiatrists to agree on what condition they were examining and treating. The *DSM-III* created multiple, distinct conditions that were based on the manifest symptoms each displayed, without reference to any underlying etiological process. Its classifications also removed the depressive, neurasthenic, and hysterical neuroses from the list of anxiety neuroses and made them separate categories of mood, dissociative, and somatization disorders, in effect, eliminating the notion of "psychoneurosis" altogether.

The anxiety category itself was carved into nine different conditions. These included three types of phobic disorders: simple phobia, social phobia, and agoraphobia, the last of which itself was split into conditions with or without panic attacks. Other categories were panic disorder, generalized anxiety disorder (GAD), obsessive-compulsive disorder (OCD), post-traumatic stress disorder (PTSD), and "conditions not otherwise specified." The rejection of any unifying etiology behind the various anxiety diagnoses thoroughly distinguished the *DSM-III* from the first two *DSM* manuals.

The developers of the *DSM-III* faced a particularly difficult challenge in how to define generalized anxiety disorder, the core anxiety condition in the earlier manuals. Following Freud, *DSM-I* and *DSM-II* used a conception of generalized anxiety as diffuse and untied to specific situations or objects.[42] The *DSM-III* similarly defined GAD as a generalized, persistent condition of at least one month's duration. More importantly, however, it stripped GAD of any precedence and relegated it to being a residual condition that was not to be diagnosed in the presence of other anxiety disorders or any non-anxiety disorders. Because GAD rarely occurs in isolation, while *DSM-III* was the reigning manual, the condition was almost never diagnosed at all. "In fact," prominent anxiety researcher David Barlow wrote, "the category of GAD in DSM-III produced so much confusion that few clinicians or investigators could agree on individuals who would meet this definition."[43] The *DSM-III* turned what had been the core anxiety disorder in the psychodynamic tradition into a virtual nonentity.

The carving up of the anxiety neuroses into separate conditions, such as OCD, social phobia, and panic disorder, had no precedent in psychiatric history. Even Freud, whose early work described the symptomatology of most of these anxiety conditions, had a unifying theory that brought them together as different manifestations of common underlying processes. The *DSM-III*, however, created the impression that distinct, clear-cut disorders existed that had little relationship to each other, to other categories in the manual, to personality dispositions, or to more generalized distress and problems in living.

Another sharp severance from dynamic conceptions was the *DSM-III*'s relegation of the anxiety disorders to a lower status than other classes of mental disorders. The decision not to diagnose anxiety disorders that were due to major depressive disorders was especially consequential. When concomitant symptoms met diagnostic criteria for both a mood disorder and an anxiety disorder, the condition would be considered a depressive, not an anxious, one. This reversed the basic tenet of *DSM-I* and *DSM-II* that anxiety, not depression, was at the core of psychoneurotic conditions. The flaw in subordinating anxiety to depressive conditions was so apparent that just seven years later *DSM-III-R* (1987) largely abandoned *DSM-III*'s hierarchical system and allowed for the diagnosis of coexisting anxiety and depressive disorders. This change, in turn, created its own major problem, namely an explosion of artificial comorbidity between seemingly separate anxious and depressive diagnoses (discussed in Chapter 8).

Before the *DSM-III* analysts generally saw anxiety as a continuum ranging from normal fears on one end, through anxious symptoms and anxiety neuroses, to disabling anxiety conditions on the other end. Similarly, psychologists, such as Hans Eysenck, portrayed various anxious states along continua of introversion to extraversion and neuroticism to stability. Eysenck mocked the new manual, asserting that it reflected the need of psychiatry to appear to be a medical specialty at the cost of ignoring overwhelming evidence that anxiety conditions were, in fact, dimensional, not categorical. Distinct categories could not distinguish between mild, moderate, and severe versions of an anxiety condition. The diagnostic specificity promulgated in *DSM III*, *III-R*, and *IV* seemed a poor fit for the anxiety disorders, which featured messy boundaries between each other and other types of disorders.[44]

THE IMPACT OF THE DSM-III

Despite its major flaws, the classificatory tradition that began with the *DSM-III* turned out to be enormously useful for psychiatry. As Chapter 6 noted, psychiatrists faced stiff competition from clinical psychologists, who had developed powerful behavioral treatments for anxiety. The *DSM-III*'s careful delineation of discrete disorders firmly tied psychiatry to the field of medicine and thus to drug-oriented treatments. This bolstered the profession's claim to special expertise in comparison to its professional competitors.

The *DSM-III* was almost immediately incorporated into psychiatric teaching and textbooks, clinical practice, research trials, public policy, and epidemiological studies. It was probably the single most significant publication in the history of psychiatry, becoming the basis for obtaining reimbursement, collecting statistical and epidemiological data, providing a basis for communication, and helping ensure that researchers were studying common entities. Psychiatric researchers and epidemiologists also benefited from the new disease categories, which were far easier to measure and study than the murky categories of the former diagnostic manuals.[45]

The National Institute of Mental Health (NIMH) was another institution that promoted the new, categorical view of anxiety. Since its inception in 1948, the NIMH had taken a broad psychosocial view of mental illnesses, including anxiety. During the late 1960s, however, its sponsorship of research that focused on social and environmental problems became politically toxic when a conservative Republican administration assumed power. By the early 1970s the psychosocial study of anxiety had become a public relations liability. The NIMH sponsored the development of the *DSM-III*, and the manual provided it with specific diagnostic entities that bolstered its claims to be concerned with the study and treatment of diseases rather than with psychosocial problems. The disease entities of *DSM-III* replaced the conceptions of anxiety that had dominated the immediate post–World War II era, which assumed that social stressors grounded in the tensions of modern life produced most anxious conditions. Diseases were far safer and more prestigious entities to study than the psychological results of stressful social environments. Claiming a mandate to sponsor research about mental illnesses diffused the political attacks on the NIMH; its budget has consistently increased since the 1970s.[46]

The epidemiological studies that the NIMH also sponsored have had an especially important impact on the public consciousness of anxiety. Past eras were more likely to view most forms of anxiety as normal reactions to fears and uncertainties. When published in 1980, the *DSM-III* noted: "It has been estimated that from 2% to 4% of the general population has at some time had a disorder that this manual would classify as an Anxiety Disorder."[47] By the early 2000s, studies indicated vastly higher rates, showing that nearly 30 percent of respondents to public surveys reported an anxiety disorder at some point in their lives. Specific phobias (the most common of which are fears of animals, heights, blood, flying, and closed spaces) and social phobias (by far the most common of which is fear of public speaking) were two of the three most prevalent particular conditions of any type.[48] These huge rates stemmed from an enormous change in thinking about the nature of anxiety. Since 1980, government agencies, the researchers they sponsor, and mental health clinicians have all benefited from using symptom-based criteria that define all anxiety conditions—regardless of whether they are natural responses or are pathological—as indicative of mental disorders.

Other institutions also profited from the *DSM-III*'s diagnostic categories. While pharmaceutical companies did not directly participate in developing *DSM-III*, they were among the major beneficiaries of its categorical system. The drug industry found the various diagnoses appealing targets for their products; they widely trumpeted the huge prevalence estimates of anxiety that epidemiological studies presumably uncovered. The categories also allowed them to meet the FDA mandate that required them to market drugs only for the treatment of recognized disease conditions. Ultimately, the number of psychiatrists employing psychopharmacology exploded while the proportion only practicing psychotherapy plunged.[49]

The anxiety disorders became especially important conditions for epidemiologists, not just because of their high prevalence but also because they apparently preceded and caused other psychiatric conditions. Most respondents in community surveys who reported a mental disorder said that an anxiety disorder was the first disorder they experienced. Social anxiety became a particular concern. "Early-onset social phobia," epidemiologist Ronald Kessler and colleagues wrote, "is a powerful predictor of the subsequent onset and course of a wide range of secondary mental and substance use disorders."[50] Researchers assumed that anxiety disorders predated and led to such adverse effects as lower educational and occupational attainment, teenage pregnancy, and divorce. Policy makers began to advocate the screening of children and adolescents for anxiety (and depressive) disorders in their schools, in order to identify and treat these conditions at early ages and, presumably, prevent more serious conditions from developing. This turned social policy toward a focus on personal factors and away from social conditions that might make people vulnerable to anxiety in the first place. After 1980, the presumed causal sequence between anxiety and social circumstances was reversed: social stressors are now more likely to be viewed as the results, rather than as the generators, of anxiety.[51]

A NEW ERA OF DRUGS

When the *DSM-III* was published in 1980, the reaction against the benzodiazepines and consequent plunge in sales had created a vacuum in the market for psychoactive drugs directed at common, distressing conditions. A new anxiolytic, Xanax, partially filled this gap. Its maker, Upjohn, promoted Xanax as a treatment for panic disorder, which first became an officially recognized anxiety disorder in the *DSM-III*. Despite intense controversy over its effectiveness and addictive potential, by 1986 Xanax had become the largest-selling anxiolytic drug. In 2009, Xanax was the single most prescribed psychiatric medication of any sort, with over 44 million prescriptions written in that year.[52]

From the late 1980s to the present, however, the class of the antidepressant selective serotonin reuptake inhibitor (SSRI) medications has overshadowed the anti-anxiety drugs. Like the benzodiazepines in the 1950s and 1960s, these drugs became hugely popular as soon as they entered the marketplace in 1988. Driven by credulous media stories and Peter Kramer's best-selling book *Listening to Prozac*, the SSRIs were initially seen as agents that could transform personalities, not just palliate mental illnesses. Their initial marketing positioned them as the "anti-Valium," which would liberate and enable but not, as the benzodiazepines presumably did, sedate and addict users. This made them, Kramer noted, feminist drugs that empowered women.[53]

Media accounts portrayed the SSRIs as analogous to "cosmetic psychopharmacology," enhancing self-esteem, sex appeal, popularity, and work performance. They widely broadcast the message that the SSRIs altered broad states of temperament, not just particular mental illnesses. "We are entering," Kramer proclaimed, "an era in which medication can be used to enhance the functioning of the normal mind."[54] One cover story in *Newsweek*, for example, marked the wide-ranging impact of the SSRI Prozac in its story title, "How Science Will Let You Change Your Personality with a Pill."[55]

The SSRIs were actually as much anxiolytic as antidepressant drugs. Kramer and others recognized that the SSRIs worked across many different conditions, including anxiety. Indeed, Kramer noted that the SSRIs "could as appropriately have been called anxiolytics."[56] The subtitle of the *Newsweek* story explicitly related the drug to many common anxious situations: "Shy? Forgetful? Anxious? Fearful? Obsessed?" Calling the SSRIs antidepressants as opposed to anxiolytics had more to do with marketing issues than the nature of the drugs themselves. The benzodiazepines—anxiolytics—had become toxic commercial products. As psychopharmacologist David Healy observed, it was "very difficult to market a general anxiolytic, in the post-benzodiazepine era."[57] Nevertheless, their labeling as antidepressants ushered in a new era: if a drug was an antidepressant, then the condition it treated must be depression, not something else. The age of anxiety was thus transformed into an age of depression.[58]

While anxiety has generally taken a back seat to depression in the cultural consciousness since 1990, there have been exceptions. Social anxiety disorder (SAD), a condition that the *DSM-I* and *II* had not mentioned, has been a major focus of drug company promotional efforts. Fueled by a massive public relations campaign costing nearly $100 million, GlaxoSmithKline promoted its SSRI, Paxil, as a remedy for the disease of "being allergic to people." "Every marketer's dream," Paxil's marketing director boasted, "is to find an unidentified or unknown market and develop it. That's what we were able to do with social anxiety disorder."[59] Paxil sales rose by nearly 20 percent during the year after the campaign began and have since reached over $2 billion.[60]

While portrayals of tranquilizers had reflected the socially centered worldview that had prevailed in psychiatry during the 1950s and 1960s, those of the SSRIs mirror an era captivated by brains, neurochemicals, and neuroscience. Advertisements for the SSRIs targeted specific diseases such as SAD and GAD rather than the diffuse states of anxiety, nerves, tension, and stress that the tranquilizers were said to allay. In addition, characterizations of the SSRIs reversed the cause and effect relationship between social conditions and psychological states. Anxiety was not the result of social stressors, as in the earlier period, but itself produced social and occupational impairments. Finally, views of chemical imbalance replaced psychosocial concepts as the reputed sources of anxiety.

In fact, Prozac and its siblings did not turn out to be "happy pills," and after an initial period of enthusiasm, reports of dramatic positive changes in personality disappeared. Indeed, in 2005, about three-quarters of patients who received a prescription for an SSRI did not refill it, suggesting that these drugs more often disappointed their users than produced miraculous transformations in them. The effectiveness of the SSRIs also did not seem to exceed that of the benzodiazepines or other earlier antidepressants. Moreover, follow-up reports associated these drugs with side effects such as weight gain, decreased sex drive, and a variety of unwanted somatic effects in some users.[61]

Nevertheless, the SSRIs became as successful as the tranquilizers had been in the 1950s and 1960s. Their sales rose from 76 million prescriptions in 1996 to 186.3 million in 2008, nearly two and a half times as many. In the same period, prescriptions of anti-anxiety medications nearly doubled, from 58.3 to 98.1 million.[62] Medication had largely replaced psychological and behavioral therapies as responses to anxiety. The seeming success of psychopharmacology also led to new therapeutic optimism and self-confidence within the psychiatric profession.[63]

CONCLUSION

Cultural forces have always influenced definitions and responses to anxiety conditions. Since the mid–1950s the nature of these forces has changed considerably. From that time until the present,

anxiety has become an aspect of commercial culture, with consumers expecting that it can be allayed through drug treatments.
During the 1950s and 1960s, however, drugs were viewed as aids
in coping with normal concerns of everyday life; since that time,
anxiety has been formulated as a variety of specific mental disorders in need of chemical correction. People are now far more likely
to make sense of their anxiety through the vocabulary of chemical
imbalances of neurotransmitters than through social, religious, or
humanistic idioms. Whether or not this reinterpretation will lead
to a greater understanding of this condition, however, remains in
question.

The more precise classification systems in *DSM-III, DSM-III-
R,* and *DSM-IV* (2000) have not yet generated greater insights
about the causes of anxiety disorders or better understandings of
how to separate natural anxiousness and neurotic temperaments
from anxiety disorders. In addition, while psychopharmacology
has largely displaced analytic and other treatments, the current
anxiolytic and antidepressant medications do not seem to out-
perform the benzodiazepines. This stalemate in acquiring better
knowledge about and treatments for anxiety and its disorders has
led to initiatives to thoroughly revise the *DSM* concepts of anxiety. In many ways these efforts resemble much earlier historical
understandings.

CHAPTER EIGHT

The Future of Anxiety

AN ANXIOUS SOCIETY

Modern developed societies are the safest, healthiest, and most prosperous that have ever existed so we might expect that their citizens would have low levels of anxiousness. "Hasn't one of the central accomplishments of modern civilization," Norwegian philosopher Lars Svendson asks, "been the overall reduction of fear, by nighttime electrical lighting, insurance policies, police forces, standing armies, the destruction of predatory animals, lightning rods on churches, solid locked doors on all buildings, and thousands of other small designs?"[1] Indeed, the daily news notwithstanding, overall levels of violence seem to be at their lowest in recorded history.[2] In addition, unprecedented longevity means that few people in developed countries need to fear dying before old age. Moreover, degrees of economic security greatly exceed those typical of eras before the postindustrial period.

Nevertheless, epidemiological surveys inform us that the public reports more anxiety disorders now than in the past. These studies indicate that anxiety is the single most common class of mental illness; almost one of five people has had an anxiety disorder during the past year and more than a quarter of the population

(28.8 percent) experienced one at some point in their lives. Moreover, these rates seem to be growing at an alarming pace: in 1980, the *DSM-III* estimated that the prevalence of anxiety disorders in the United States was only between 2 and 4 percent.[3]

Given the general decline in dangerous situations and increase in health, safety, and material well-being, it does not seem plausible that the upsurge in measured rates of anxiety represents a genuine epidemic of anxiety disorders. Instead, the putative growth of anxiety disorders since the implementation of the *DSM-III* seems to stem primarily from the use of diagnostic criteria that fail to accurately distinguish natural from disordered anxiety. From Aristotle to Burton to Freud, diagnosticians separated anxious conditions that naturally emerged in response to external dangers from those that were not rooted in genuine threats, calling only the latter disorders. Since 1980, however, psychiatry has primarily used symptom clusters to define disorders, without referring to the reasons symptoms emerged. This practice means that all anxiety conditions of great enough severity are diagnosed as disorders, regardless of whether they are present for good reason or not.

The large and increasing rates of anxiety disorders are one result of the tremendously successful implementation of the *DSM* manuals since 1980. These manuals followed the traditional medical model of well-defined, specific entities in their diagnostic criteria; and the widespread adoption of the *DSM* categories by clinicians, insurers, drug companies, researchers, policy makers, and mental health advocates validated psychiatry's claims to special expertise in diagnosing and treating mental illnesses. Another consequence of using the medical model was that the *DSM* diagnostic categories generated large estimates of anxiety disorders, which in turn created more clients for clinicians, bigger markets for the pharmaceutical industry, and a "public health problem" that validated the mission of mental health policy makers, researchers, and advocates. The current impression of an "epidemic" is sustained by the many groups that benefit from identifying and treating anxiety disorders but have nothing to gain from considering anxious symptoms as normal.

To be sure, psychiatry and other interested groups rarely impose disease labels on a resistant public. A general cultural inclination to view distressing emotions as disorders has grown rapidly in recent years, lowering the threshold for when people accept psychiatric diagnoses. The *DSM* created the kind of illness conditions that allowed patients to obtain compensated treatments for their distress. The manual's medical idioms have generally replaced social, religious, and philosophical understandings of the nature of anxiety as people have become more willing to define psychic and bodily signs of anxiety as symptoms of diseases that require medication or other kinds of professional therapies.[4] By the beginning of the twenty-first century, the *DSM* was thoroughly institutionalized among the general public as well as in all aspects of the mental health enterprise.

ANXIOUS BRAINS

Despite the manual's tremendous practical success and cultural acceptance, a gaping hole existed in the *DSM*'s theoretical edifice. A fundamental principle of the *DSM-III* revolution was that diagnostic criteria could not use etiology to define conditions. Since 1980, however, biological understandings have become foundational for psychiatric research and treatment. Anxiety disorders (among others) are now viewed as brain-based conditions rooted in neurotransmitters, neural networks, and genes.

This biological paradigm considers all sorts of anxious experiences to be organic phenomena. "Perhaps," psychiatrist Peter Kramer asserts, "what Camus's Stranger suffered—his anhedonia, his sense of anomie—was a disorder of serotonin."[5] The opening statement of neuroscientist Joseph LeDoux's popular *Synaptic Self: How Our Brains Become Who We Are* boldly states: "The bottom-line point of this book is 'You are your synapses.'"[6] For LeDoux and other neuroscientists, anxiety is encoded in brain circuitry and stored in synaptic changes. Accordingly, researchers' major task is to locate the underlying brain mechanisms that both lead to anxiety and must be changed to control it.

The brain-based paradigm is now widely regarded as the sole

basis of therapeutic legitimacy. Even approaches that had been militantly nonsomatic in the past now justify their methods through their ability to change brains. For example, responding to an article about psychoactive medication, the president of the American Academy of Psychoanalysis and Dynamic Psychiatry complained that the author did not realize that psychoanalysis also works through affecting brains: "I encourage him to review the emerging medical literature on how psychotherapy causes measurable structural and functional metabolic changes in the brain."[7] Another psychotherapist introduced her book with the claim: "Let me say at the outset that in calling psychotherapy neurosurgery, I am not speaking metaphorically. We now have solid scientific evidence to suggest that the so-called 'talking cure,' originally devised by Freud, literally alters the way in which the neurons in the brain are connected to one another."[8] Cognitive therapists, too, assert that their techniques change brains in the same ways drugs do.[9] Credible professional practice is coming to depend on tying treatments to their impact on the brain.

Moreover, neuroscientific understandings have moved beyond practitioners to penetrate the general culture. Just as earlier forms of discourse framed anxiousness in terms like "neurasthenia," "neurosis," or "stress," people now use their imagined neurochemical levels to interpret everyday experiences. For example, a *New Yorker* essay describes a typical interaction between Paul and Patricia Churchland, two prominent philosophers:

> One afternoon recently, Paul says, he was home making dinner when Pat burst in the door, having come straight from a frustrating faculty meeting. She said: "Paul, don't speak to me, my serotonin levels have hit bottom, my brain is awash in glutocortocoids, my blood vessels are full of adrenalin, and if it weren't for my endogenous opiates, I'd have driven the car into a tree on the way home. My dopamine levels need lifting. Pour me a Chardonnay and I'll be down in a minute." Paul and Pat have noticed that it is not just they who talk this way—their students now talk of psychopharmacology as comfortably as they talk of food.[10]

Not just philosophers are talking this way. "Zoloft should send a comet into this place," an observer of the anxiety of fashion show models during the height of an economic crisis exclaims, "So should Paxil. They should shower this place with pills. Serotonin levels are down."[11] A cultural discourse has emerged that assumes anxiety reflects biological rather than psychological, existential, or sociological properties.

This recent thoroughgoing grounding of all forms of anxiousness in neurochemistry rivals the most ambitious historical attempts to somaticize anxiety. In an 1835 publication, phrenologist Joseph Gall observed, "Whoever would establish . . . on the intellectual and moral functions . . . a solid doctrine of mental diseases, of the general and governing influence of the brain in the states of health and disease, should know . . . that it is indispensable . . . that the study of the organization of the brain should march side by side with that of its functions."[12] This statement would not look out of place on a current APA or NIMH website. Yet, perhaps the most curious aspect of the current grounding of anxiety disorders and their treatments in brain functioning and the central nervous system is the total absence of such understandings in psychiatry's current foundational document, the *Diagnostic and Statistical Manual*.

A MAJOR UPDATE FOR THE DSM

Paradoxically, psychiatry's diagnostic manual has remained completely immune to the field's biological revolution. The development of neuroimaging tools that revealed the neural circuitry associated with anxiety and the mapping of the genome seemed to have the potential to lead to a new era for discovering the causes of and treatments for anxiety disorders. In particular, the Human Genome Project, which was completed in 2003, promised to advance understandings about the molecular and genetic underpinnings of anxiety and other mental disorders. Yet, the extant diagnostic criteria did not take these achievements into account.

To banish psychodynamic influences, the developers of the *DSM-III* had insisted that the manual be militantly atheoretical.

The core *DSM* principles, which use constellations of symptoms to define each discrete disorder, separate the class of anxiety disorders from other classes of disorders, treat each anxiety condition as distinct from the others, and starkly distinguish normal from disordered conditions, have remained unaltered since the revision of the third edition in 1980. While the *DSM-III-R* and *DSM-IV* changed the hierarchical structure of diagnoses and the particular criteria of some of the anxiety disorders, they did not incorporate biological understandings in any of their definitions of anxiety (or other) disorders. The doctrine that etiology could not be used to define particular disorders endured. The result was that the diagnostic manual did not incorporate neuroscientific advances in its definitions.

While general agreement existed that the manuals since the *DSM-III* had created more reliable diagnostic criteria, few thought that they had solved the problem of how to define categories that are more valid, and many even felt that they posed barriers to untangling the causes, prognoses, and treatments for the various anxiety conditions. The findings from brain-based studies did not conform to the *DSM*'s categorical model but were more compatible with the view that the various anxiety disorders share features with each another, with mood disorders, and with normal temperamental variation. In particular, the sharp *DSM* boundaries between natural anxiousness and anxiety disorders poorly fit newly discovered information about the nature of anxiety conditions. Psychiatrist Jordan Smoller acknowledges: "Focusing on the abnormal has led to a system of classification—the DSM—based on constructing categories from constellations of symptoms. Without a map of how these symptoms connect to the functional organization of the mind and brain, it's hard to evaluate their validity."[13]

The issue of validity takes on particular importance for psychiatrists because, unlike most physical conditions, for which biological markers exist that can confirm or refute diagnoses, the *DSM* criteria are the only resource with which they can detect pathology. Therefore, psychiatry's diagnostic criteria have an outsized place in the field compared to those of other medical specialties.

Researchers who needed to distinguish normal from disordered anxiety could only compare groups that were predefined by the highly flawed *DSM* criteria. These concerns motivated the American Psychiatric Association to undertake a major revision of the diagnostic manual.

Despite the thorough institutionalization of the *DSM* in teaching, clinical practice, research, treatment outcomes, and insurance reimbursement and the absence of any competing diagnostic system,[14] in 1999 the APA, in cooperation with the NIMH and World Health Organization, initiated what it envisioned would be an important revision of the manual. As they had for all previous editions of the manual since the *DSM-III*, psychiatric researchers controlled the *DSM-5*[15] process. Their major assumption was that the vast progress made in neuroscientific knowledge of brain functioning since the last significant alteration of the *DSM* could inform a new edition of the diagnostic manual. "As new findings from neuroscience, imaging, genetics and studies of clinical course and treatment response emerge," declared the *DSM-5* task force, "the definitions and boundaries of disorders will change."[16]

Most of the attention surrounding the development of the *DSM-5*, which was released in May 2013, centered on the bitter and highly public disputes that marked its development, rather than the scientific issues that underlay the revisions. Some of these arguments were procedural. The initial deliberations regarding the revision were shrouded in secrecy. The APA had required each member of the work groups to fill out nondisclosure forms, insuring that their discussions would be private. Not until 2010, when Robert Spitzer and Allen Frances, the chief editors of the *DSM-III* and *DSM-IV*, respectively, made vigorous and highly public protests, did the proposals of these committees become open for public comment. Spitzer was "dumbfounded" when Darrel Regier, the APA's director of research and vice chair of the task force, refused his request to see the minutes of meetings of the *DSM-5* task force. Soon thereafter, he was again appalled to discover that the APA had required psychiatrists involved with the revision to sign a paper promising that they would never talk about what they were doing, except when necessary for their jobs. "The intent," according to Spitzer, "seemed to be not to let anyone know what the hell was going on."[17]

Allen Frances, the chief editor of the *DSM-IV*, provided perhaps the harshest assessment of the deliberations leading up to the new manual. Frances accused "his colleagues not just of bad science but of bad faith, hubris, and blindness, of making diseases out of everyday suffering and, as a result, padding the bottom lines of drug companies."[18] The APA's official response to Frances was to accuse him (and Spitzer) of "pride of authorship" and to point out that his royalty payments would end once the new edition was published—a fact that "should be considered when evaluating his critique and its timing."[19] What was supposed to be a deliberative, scientific process had turned into a schoolyard brawl. While media attention focused on these colorful controversies, the *DSM-5* work group for the anxiety disorders produced a number of significant substantive changes.

DSM-5 PROPOSALS FOR ANXIETY DISORDERS

In 2007, the APA appointed a variety of work groups to develop *DSM-5* criteria for specific categories of disorders. The anxiety disorders work group recommended three major changes: dimensionalizing the anxiety criteria, establishing a mixed anxiety/depressive diagnosis, and substantially altering the generalized anxiety disorder (GAD) criteria.[20] Each of these changes reflected significant historical themes.

Dimensionalization

The psychodynamic conception that dominated views of anxiety before the *DSM-III* considered mental disorders to vary along a dimension of lesser to greater severity. Anxiety conditions formed a continuum, with normal, contextually appropriate anxiety on one end and severe anxiety disorders on the other. This conception of graduation differed from the medical model of disease classification, which usually featured discrete disease entities that had sharp boundaries separating them from healthy states. It also diverged from Kraepelinian approaches to psychiatric diagnosis, which the categorical and symptom-based Feighner criteria had restored as a prominent strategy among psychiatric researchers.

One of the major motives behind the *DSM-III* was to establish the kind of distinct conditions that prevailed in other medical classifications, yet the sharp cut-offs between normal and disordered conditions that the diagnostic criteria imposed had little scientific justification. The diagnostic system established by the *DSM-III* seemed to many to be both arbitrary and unable to recognize minor forms of disorder. For example, there was no logical reason why GAD diagnoses required at least three symptoms or panic attacks at least four symptoms: people with fewer symptoms might simply have milder cases of the disorder. Moreover, the *DSM*'s binary logic conflicted with understandings of the subtler ways in which biological and genetic variance became manifest. A dimensional system might provide a better fit than a categorical one for the underlying nature of anxiety conditions, which has no accepted cut-off point for where the number of symptoms distinguishes disorder from nondisorder. The either/or logic of the recent manuals seemed to hinder etiological discoveries.[21]

The *DSM 5* task force proposed major alterations that would move the manual from a categorical toward a more dimensional system. Initially, it suggested a radical revision that would largely have replaced the distinct criteria dividing disorders from nondisorders with measures that reflected graded scales of severity:

> The single most important precondition for moving forward to improve the clinical and scientific utility of DSM-V will be the incorporation of simple dimensional measures for assessing syndromes within broad diagnostic categories and supraordinate dimensions that cross current diagnostic boundaries. Thus, we have decided that one, if not the major, difference between DSM-IV and DSM-V will be the more prominent use of dimensional measures in DSM-V.[22]

Dimensions, the revisers hoped, would overcome the inability of categorical measures to identify people who show some, but not enough, symptoms to qualify for a diagnosis.

The developers of the *DSM-5* recognized, however, that their original goal of a dimensional system of measurement was overly ambitious and, given the existing knowledge base, premature. They scaled back their proposal to a suggestion that clinicians gather dimensional measures to supplement, not to replace, the categorical criteria: "When an initial screening reveals one or more symptoms in different domains, patients and clinicians will be directed to follow up with a more intensive dimensional evaluation of symptom severity, followed by a clinical judgment about whether the symptoms are sufficient for a mental disorder."[23] The scale they proposed for anxiety asks respondents whether they "have felt" the following symptoms over the previous seven days: fearful, anxious, worried, hard to focus on anything except my anxiety, nervous, uneasy, tense. The numerically scaled answers— 1 = Not at all, 2 = A little bit, 3 = Somewhat, 4 = Quite a bit, 5 = Very much—would be summed and the total checked against a continuous scale of severity.[24] This scale presumably would allow clinicians to consider degrees of severity within a diagnosis as well as to identify patients who did not have a full-blown condition.

Dimensional assessment invoked a measurement style from an era that preceded the *DSM-III*. During the 1950s and 1960s clinicians and epidemiologists commonly assumed that a small number of anxious symptoms often indicated a milder form of anxiety disorder, not the absence of disorder. Epidemiological surveys at the time, which used continuous scales, uncovered immense rates of graded anxiety-based conditions. The best-known survey, the Midtown Manhattan Study, found that just 18.5 percent of its community sample was symptom free.[25] Over 80 percent of the population, therefore, had some degree of "mental illness": 36 percent were in the mild category, 22 percent were in the moderate category, and 23 percent fell into the marked, severe, or incapacitated category. The findings showing that a large majority of people had some degree of mental illness were widely mocked at the time. Indeed, one of the reasons for developing the categorical conditions of the *DSM-III* was to avoid pathologizing such a huge proportion of the population. The *DSM-5* dimensional diagnostic criteria, which use ubiquitous symptoms such as feeling fearful, anxious, or worried that need endure for such a brief period of time as seven days, would likely produce similarly massive rates of "anxiety disorders," although many could be classified as mild.

The *DSM-5* task force eventually abandoned its dimensional proposal. The reason, however, was not because of fears of overpathologizing anxiety and other conditions. Instead, the divergent needs of researchers and clinicians led to the rejection of the task force's proposal. While researchers were most interested in developing diagnoses that would improve psychiatry's knowledge base, clinicians had the more practical concerns that the criteria be easy to apply and guarantee reimbursement for treatment. Clinicians worried that dimensions would be burdensome to use in practice, especially if insurance companies mandated their employment. A meeting of the APA Assembly rejected the dimensional proposal, in effect, by voting to retain the current categorical diagnoses. A proposal that began as a radical revision of the basic structure of the *DSM* ended as an appendix offering a suggestion in need of further study.

Ironically, the empirically driven researchers had imposed the *DSM-III*'s categorical system on resistant clinicians, yet by 2012 the categorical diagnoses had become millstones for researchers but necessities for clinicians, who required them to be paid for treatment. Clinicians thus received a measure of revenge on researchers by rejecting their appeal to institute a possibly more valid, but less practically useful, diagnostic system. Social considerations mandated that psychiatry continue to employ a strict, nondimensional classification system whose scientific inadequacies had become blatantly obvious.

Mixed Anxiety/Depression

A fundamental problem of the *DSMs* beginning in 1980 was the immense amount of actual comorbidity between supposedly distinct conditions. As such broad concepts such as "melancholia," "neurasthenia," and "psychoneurosis" had recognized, anxiety rarely appears in pure forms. In 1621 Robert Burton had emphasized that fear and sorrow were "continual companions."[26] Well before Burton, the Hippocratics had joined fear and sadness at the hip: "when fear and sadness last for a long time, this is melancholy."[27] Psychiatric history, through the *DSM-II*'s categorizing of depressive neurosis as a displacement of anxiety, perennially regarded depression and anxiety as integrally connected.

The *DSM-III*'s pointed separation of anxiety and depression, which has persisted in each version of the manual since 1980, was a historical anomaly. Not surprisingly, empirical research indicated that the vast majority of anxiety states were comingled in various combinations with depressive, psychosomatic, phobic, obsessive-compulsive, and other syndromes. In contrast to most medical diagnoses, in which borderline cases that fall between categories are rare, few people received a diagnosis of a single anxiety condition without receiving diagnoses of additional anxiety disorders. Also, mixed cases of anxiety and mood disorders were more of a rule than an exception: anxious persons generally presented concurrent symptoms of depression, and depressed persons showed symptoms of anxiety. All of the anxiety disorders (with the possible exception of obsessive-compulsive disorder) were strongly related to mood disorders; in epidemiological studies, scales that measured depression and anxiety typically overlapped by about 60 percent. Especially when temporal patterns are considered, constant states of a single diagnosis were exceptional.[28]

The comorbidity of anxiety and depressive disorders was comparable in patient populations where about two-thirds receive both diagnoses over their lifetimes. Moreover, the vast majority of anxiety states do not remain static but transition to depressive and other diagnoses over time. "Only a small proportion of individuals exhibit 'pure' forms of [anxiety] conditions cross-sectionally, and even fewer across the life course," concluded an international task force in 1996.[29] In addition, research has revealed that common rather than discrete genetic and personality factors seem to underlie anxiety and depression as well as the various anxiety conditions.[30] For example, the same genes influence the risk of developing the various anxiety disorders and major depression. Genes apparently transmit a more general trait related to neuroticism, not to any particular condition.

All of these factors notwithstanding, the logical structure of the *DSM* mandated that anxiety and depression be separate diagnoses whose conjunction indicated the coexistence of two distinct conditions. The categorical classification system that began with the *DSM-III* ignored the tremendous inherent overlap among the various anxiety disorders and between anxiety and depression. In 1997 psychopharmacologist David Healy, surveying the hundreds of discrete diagnoses in the *DSM-IV*, presciently noted, "It would seem inevitable that there must be a collapse back toward larger disease categories at some point." A few years earlier, psychiatrist Peter Kramer asserted, "A good case can be made for the return of 'neurosis,' a catchall category for serious minor discomfort related to depression and anxiety."[31]

The *DSM-5* anxiety work group had to deal with this glaring problem of validity in the existing diagnostic criteria. "How then are we to update our classification," the major developers of the *DSM-5* wondered, "to recognize the most prominent syndromes that are actually present in nature, rather than in the heuristic and anachronistic pure types of previous scientific eras?"[32] The phrase "anachronistic pure types of previous scientific eras" referred to the *DSM-III*'s discrete diagnostic categories. The creators of the *DSM-5* did not acknowledge that the sharp split of anxiety and depression that began in the *DSM-III* was actually a distinct break with previous psychiatric history.

The anxiety disorders work group initially proposed to deal with the problem that anxiety and depression rarely occur in isolation through introducing a new diagnosis of mixed anxiety/depression. This diagnosis would apply to patients who had three of four depressive symptoms accompanied by two symptoms of anxious distress (irrational or preoccupying worries, trouble relaxing, tension, catastrophic fears), but who did not qualify for another anxiety or depressive diagnosis. The symptoms would need to have lasted for only two weeks, far shorter than the duration required for other anxiety diagnoses.

The work group came up with the mixed anxiety/depression diagnosis as the result of empirical findings showing a huge co-morbidity between the two conditions. Yet, what they presented as a scientific advance stimulated by empirical research in fact represented a return to the sort of less differentiated condition that had dominated psychiatric thought for millennia. The work group's proposal for an anxiety/depression diagnosis would have reestablished the perennial connection between the two conditions that was well-recognized before 1980.

While the new diagnosis might have helped to resolve the problem of comorbidity, it did so in a way that pathologized a new, and potentially vast, category of people. The proposed diagnosis contained no way of separating acute, stress-related symptoms from true anxiety disorders. Given the ubiquity of such symptoms, coupled with the minimal number and duration of symptoms required for the diagnosis, the new criteria were likely to result in high rates of false positive diagnoses. A number of experts criticized the criteria because of their low threshold and inability to distinguish disordered from normally distressing anxiety and depression. These experts' very public protests against labeling ordinary distress a mental disorder were strong enough that the APA withdrew the proposal shortly before the manual was completed.[33] The result is that the *DSM-5* will continue the separation of depression from anxiety that the *DSM-III* launched.

Generalized Anxiety Disorder

The *DSM-5* anxiety work group also proposed changes to particular criteria for the diagnosis of various anxiety disorders. The most important of these changes concern generalized anxiety disorder (GAD). The *DSM-I* and *DSM-II*, following Freud, defined anxiety reactions as anxious expectation or overconcern that was "diffuse and not restricted to definite situations or objects."[34] Similarly, the *DSM-III* considered GAD to be a "generalized, persistent" condition that lacked the more specific symptoms characterizing the other anxiety disorders. It also made GAD a residual condition that couldn't be diagnosed in the presence of other anxiety or depressive diagnoses.[35] Because most patients who met the GAD criteria also met the criteria for other disorders, the actual number of GAD diagnoses was quite low.

The *DSM-III-R* (1987) abandoned the hierarchical rule that disallowed GAD diagnoses in the presence of other disorders. It also changed the description of the nature of GAD from generalized anxiety to a worry-based condition, defining it as: "Unrealistic or excessive anxiety and worry (apprehensive expectation) about two or more life circumstances, e.g. worry about possible misfortune to one's child (who is in no danger) and worry about finances (for no good reason), for a period of six months or longer, during which the person has been bothered more days than not by these concerns."[36] The central place this definition accorded to worries not only changed the core nature of the diagnosis but also—given the omnipresence of things that people have to worry about—potentially pathologized many common anxious feelings. However, the many qualifiers in the definition, such as "two or more life circumstances" and references to anxiety about children who are "in no danger" or about finances "for no good reason" clearly distinguished realistic worries from anxiety disorders. The fact that the *DSM-III-R* also required that symptoms endure for at least six months reduced the possibility that situational anxiety would be misdiagnosed as GAD.

The *DSM-IV* abandoned the contextualization of the previous manual such that even anxiety over activities like work or school performance that endured for six months was diagnosable. The *DSM-5* continues to relate GAD to realistic (albeit "excessive") worries, stating that GAD involves "excessive anxiety and worry (apprehensive expectation), occurring more days than not for at least six months, about a number of events or activities (such as work or school performance)."[37] The diagnostic criteria of the two most recent *DSM*s fundamentally change GAD from a condition that is unrelated to actual concerns to one that reproduces the most common genuine bases for anxiousness ("such as work or school performance"). Because the definition doesn't clarify the meaning of "excessive" (e.g., patient self-definition, social normals, clinician judgment), this term does not provide much help in limiting false positive diagnoses. The current characterization of GAD, unlike the criteria that predated the *DSM-IV,* fails to distinguish anxiety that is unattached to realistic fears from contextually appropriate worries.

In certain respects, the *DSM-5* recommendations for dimensional measurement and a mixed anxiety/depression diagnosis evoke the undifferentiated and continuous conceptions of anxiety that dominated most of psychiatric history. The GAD criteria, however, reverse the ages-old dictum that mental disorders must be "without cause." From the Hippocratics' separation of anxious symptoms that were with cause from those that were without, through the realistic worries of Richard Napier's clients, to the injunction in *DSM-I* and *II* that anxiety disorder "must be distinguished from normal apprehension or fear, which occurs in realistically dangerous situations," concerns about family, health, finances, or work would have seemed natural, not disordered.[38] Indeed, the GAD proposal pathologizes the most obvious and understandable sources of normal anxiety, turning a ubiquitous aspect of the human condition into a billable illness.[39] If future studies use the new GAD criteria, the number of anxiety disorders in the community will continue their exponential increase. Such growth will not indicate any actual increase in the number of sufferers from an anxiety disorder; instead it will reflect how socially influenced diagnostic criteria determine when anxiety is thought of as normal and when it is considered disordered.

CONCLUSION

The inherent blurriness of boundaries between natural and abnormal anxiety gives different societies much leeway in where to draw the lines among the various anxiety disorders, between anxiety and other conditions, and between normal and pathological anxiety. While our nature as humans seems to doom us to be an anxious species, our cultures provide the criteria that judge who among the anxious are normal and who disordered.

The categorical revolution of the *DSM-III* promised much. Certainly, the manual fit psychiatry's professional needs at the time. It bolstered the profession's legitimacy as a medical discipline and became the standard language about mental illness for not just the mental health professions but also the culture at large. In addition, it became deeply embedded in all aspects of mental health practice: clinicians and hospitals depend on its diagnoses to obtain reimbursement, insurers rely on its categories to fund treatment, and public health statistics use its classifications. Yet, the original promise of the *DSM-III* in 1980 was also that a clear, precise, and reliable diagnostic system would eventually lead to more accurate knowledge about the causes, prognoses, and treatments of mental disorders. This hope has not been realized. In particular, symptom-based categorical diagnoses, however useful they might be for psychiatric practice, poorly fit anxiety conditions. Common genetic and other influences have been demonstrated to underlie multiple expressions of anxiety; symptom constellations display considerable overlap; the same treatments can apply across diverse categories; and lines between natural and disordered states of anxiousness are in reality not sharply distinct.

A particular problem revealed by recent research is that normal, no less than pathological, anxiety is in some sense located in the brain. While technological advances can map anxiety states with amazing precision, they cannot yet distinguish those neural networks and genes related to natural anxiety from those that are associated with anxiety disorders. Neuroscientific studies have been forced to rely on the highly flawed *DSM* distinctions to separate normal from pathological anxiousness, and they have proven inadequate. Yet, the *DSM*'s great practical success has made nearly impossible the introduction of major changes in the manual that would provide a better guide for researchers. Despite the categorical system's conspicuous scientific inadequacies and the obstacles it poses to advancing brain-based inquiries, the manual's thorough institutionalization has thwarted attempts to comprehensively revise it.[40]

The intrinsic embedding of conceptions of anxiety in social, as well as natural, processes prevents even the most advanced classifications and technologies from definitively separating normal from pathological anxiety. While anxiety is a biological process, anxiety disorders are socially constructed on the basis of the prevailing circumstances in any particular time. In the twenty-first century, anxiety has become a highly valued commodity among clinicians, drug companies, epidemiologists, policy makers, and mental health advocates, yet its value rests on classifying anxious experiences, which may be nonpathological, as anxiety disorders. Because the *DSM* has become so widely accepted as authoritative, only conditions that it recognizes as disorders can be eligible for treatment cost reimbursement, become the legitimate targets of psychoactive drugs, or serve as the objects of public health campaigns, epidemiological surveys, or advocacy efforts. The unprecedented rates of anxiety disorders in the modern world reflect social interests that shape conceptions of anxiety in terms of molecules, neural networks, neurotransmitters, and genes as much as social factors once influenced conceptions of melancholia, nerves, neurasthenia, and neuroses.

The dilemmas faced in the *DSM-5*'s attempts to accommodate current neuroscience reflect ancient and perennial themes in the history of anxiety. Where should societies place the boundary between normal fears and pathological anxiety? Is anxiety a specific syndrome or an aspect of a more generalized condition? How are anxiety and depression related? Are anxiety conditions discrete or parts of a continuum? When do anxiety symptoms reflect temperament and when do they signify an underlying disorder? Despite the tremendous advances neuroscientists have made in understanding the structure and function of the brain and in developing new technologies to reveal brain processes, psychiatry has made little progress in providing answers to these questions. "The line separating healthy from pathological," psychiatric historian Janet Oppenheim notes, "is not sharper to psychiatrists now than it appeared to their Victorian forefathers, and political, ideological, or cultural biases are no less potent in defining normalcy and its opposite."[41] If the development of the *DSM-5* provides a good road map, the perennial uncertainties and social influences that have characterized past knowledge about anxiety will continue to infuse future understandings of this universal experience.

NOTES

Chapter 1. Afraid

1. Good general accounts of modern views of anxiety are found in: LeDoux J (1996). *The Emotional Brain: The Mysterious Underpinnings of Emotional Life*. Simon & Shuster, New York. LeDoux J (2002). *Synaptic Self: How Our Brains Become Who We Are*. Penguin, New York. Damasio A (1994). *Descartes' Error: Emotion, Reason, and the Human Brain*. HarperCollins, New York. Smoller J (2012). *The Other Side of Normal: How Biology Is Providing the Clues to Unlock the Secrets of Normal and Abnormal Behavior*. William Morrow, New York.

2. The current edition of this work, soon to be succeeded by the fifth edition, is American Psychiatric Association (2000). *Diagnostic and Statistical Manual of Mental Disorders,* 4th ed., text rev. American Psychiatric Association, Washington, DC.

3. Substance Abuse and Mental Health Services Administration (2012). *Mental Health, United States, 2012*. HHS Publication No. 12-4681. Department of Health and Human Services, Rockville, MD, pp. 138–139. Mojtabai R, Olfson M (2010). National trends in psychotropic medication polypharmacy in office-based psychiatry. *Archives of General Psychiatry* 67, pp. 26–36.

4. Shorter E (1992). *From Paralysis to Fatigue: A History of Psychosomatic Illness in the Modern Era*. Free Press, New York, p. 320. Shorter's notion of a "cultural symptom pool" is consistent with the difference between genotypes, which are biologically fixed, and phenotypes, which are shaped by culture, among other influences. The study of epigenetics, the branch of genetics that deals with how external influences affect gene expressions, has also become a major area of recent concern.

5. Da Costa JM (1871). On irritable heart: A clinical study of a form of functional cardiac disorder and its consequences. *American Journal of the Medical Sciences* 61, pp. 17–52. Shephard B (2001). *A War of Nerves: Soldiers and Psychiatrists in the Twentieth Century*. Harvard University Press, Cambridge. Jones E, Vermaas RH, McCartney H, Beech C, Palmer I, Hyams K, Wessely S (2003). Flashbacks and post-traumatic stress disorder: The genesis of a 20th-century diagnosis. *British Journal of Psychiatry* 182, pp. 158–163.

6. Nemiah JC (1985). Anxiety states (anxiety neuroses). In Kaplan HI, Sadlock BJ (eds.), *Comprehensive Textbook of Psychiatry*. Williams & Wilkins, Baltimore, pp. 883–894.

7. Glas G (2003). A conceptual history of anxiety and depression. In den Boer

JA, Sitsen JMA (eds.) *Handbook on Anxiety and Depression: A Biological Approach,* 2nd ed. Marcel Dekker, New York, pp. 1–48.

8. Burton R (1621/2001). *Anatomy of Melancholy.* (Jackson H, ed.). New York Review Books, New York, p. 431.

9. Rachman SJ (1990). *Fear and Courage.* 2nd ed. WH Freeman, New York.

10. Kierkegaard S (1844/1980). *The Concept of Anxiety* (Thomte R, trans.). Princeton University Press, Princeton, p. 170.

11. See especially Shorter E (1994). *From the Mind Into the Body.* Free Press, New York. Kleinman A (1988). *Rethinking Psychiatry: From Cultural Category to Personal Experience.* Free Press, New York.

12. Auster P (2012). *Winter Journal.* Faber & Faber, London, p. 141.

13. E.g., Smoller J (2012). *Other Side of Normal.*

14. Insel TR (2009). Translating scientific opportunities into public health impact: A strategic plan for research on mental illness. *Archives of General Psychiatry* 66, pp. 128–133.

15. Quoted in Robinson DN (1995). *An Intellectual History of Psychology,* 3rd ed. University of Wisconsin Press, Madison, p. 275.

16. Quoted in Porter R (2002), *Madness: A Brief History.* Oxford University Press, New York, p. 37. The meaning of the ancient Greek word for "brain" is broader than that in modern English usage. It includes understanding and sensibility as well as corporeal presence.

17. Aristotle (1980). *The Nichomachean Ethics* (Ross D, trans.). Oxford World Classics, New York, p. 127. Aristotle's term "brutish" refers to bad traits that people have by their natures.

18. Burton 1621/2001, vol. I, p. 145.

19. Krieder T (2012). Cycle of fear. *New York Times,* May 7. *http://opinionator .blogs.nytimes.com/2012/05/07/fear-and-cycling/.*

20. Kagan J, Reznick S, Snidman N (1988). Biological basis of childhood shyness. *Science* 240, pp. 167–171. Kagan J, Snidman N (2004). *The Long Shadow of Temperament.* Belknap Press, Cambridge, MA.

21. Cheyne G (1733/1991). *The English Malady* (Porter R, ed.). Tavistock/Routledge, New York. Napier quoted in MacDonald M (1983). *Mystical Bedlam: Magic, Anxiety, and Healing in Seventeenth-Century England.* Cambridge University Press, London, p. 67. Baer JC, Kim M, Wilkenfeld B (2012). Is it Generalized Anxiety Disorder or poverty? An examination of poor mothers and their children. *Journal of Child and Adolescent Social Work,* 29, pp. 345–355, quote on p. 354. Watson JB (1924). *Behaviorism.* People's Institute, Chicago. Mirowsky J, Ross C (2003). *Social Causes of Psychological Distress,* 2nd ed. Aldine de Gruyter, New York.

22. E.g., Hippocrates (1923–1931). *Works of Hippocrates* (Jones WHS, Withington ET, eds. and trans.). Harvard University Press, Cambridge. Burton, 1621/2001. Meyer A (1950–1952). *The Collected Papers of Adolf Meyer.* Johns Hopkins University Press, Baltimore. Rosenberg CE (2007). *Our Present Complaint: American Medicine, Then and Now.* Johns Hopkins University Press, Baltimore.

23. Burton, 1621/2001, vol. II, p. 245.

24. Rosenberg, 2007, p. 118.

25. Psalm 46:1–3.

26. Quoted in Y-F Tuan (1979). *Landscapes of Fear*. University of Minnesota Press, Minneapolis, p. 77.

27. See, e.g., Malinowski B (1948). *Magic, Science and Religion and Other Essays*. Free Press, Glencoe.

28. Marino G (2012). The doctor of dread. *New York Times*. March 18, p. E8.

29. E.g., Beck JS, Beck AT (2011). *Cognitive Behavioral Therapy*, 2nd ed. Guilford Press, New York. Graver MR (2007). *Stoicism and Emotion*. University of Chicago Press, Chicago.

30. Clark MJ (1995). Anxiety disorders: Social section. In Berrios GE, Porter R (eds.), *A History of Clinical Psychiatry*. New York University Press, New York, pp. 563–572.

31. Aristotle, *Nicomachean Ethics*, p. 30.

32. Kramer SN (1959). *History Begins at Sumer: Thirty-Nine Firsts in Recorded History*. University of Pennsylvania Press, Philadelphia, p. 224.

33. Freud S (1936/1963). *The Problem of Anxiety*. WW Norton, New York, p. 113.

34. The distinction between normal and abnormal sources of fear is especially complicated because the most common sorts of fears, such as heights, darkness, and crawling animals like snakes and spiders, might be evolutionarily natural fears. See Horwitz AV, Wakefield JC (2012). *All We Have to Fear: Psychiatry's Transformation of Natural Anxieties into Mental Disorders*. Oxford University Press, New York.

35. Burton, 1621/2001, p. 173.

36. Erasmus D (1974). *The Correspondence of Erasmus: Letters 1 to 141, 1484–1500*. University of Toronto Press, Toronto, p 115

37. Borsh-Jacobsen M (2009). *Making Minds and Madness: From Hysteria to Depression*. Cambridge University Press, New York.

38. Kessler RC, Chiu WT, Demler O, Walters EE (2004). Prevalence, severity, and comorbidity of 12-month DSM-IV disorders in the National Comorbidity Survey Replication. *Archives of General Psychiatry* 62, pp. 617–627.

39. Micale M (1995). *Approaching Hysteria: Disease and Its Interpretations*. Princeton University Press, Princeton, p. 120.

40. Oatley K (1992). *Best Laid Schemes: The Psychology of Emotions*. Cambridge University Press, Cambridge, p. 55.

41. Fisher P (2002). *The Vehement Passions*. Princeton University Press, Princeton, p. 157. Roberts RC (2003). *Emotions: An Essay in Aid of Moral Psychology*. Cambridge University Press, Cambridge, p. 190. Griffiths PE (1997). *What Emotions Really Are: The Problem of Psychological Categories*. University of Chicago Press, Chicago, p. 55.

Chapter 2. Classical Anxiety

1. Although the two major figures in Greek medicine, Hippocrates (460–377 BCE) and Galen (130–201 CE), both discuss anxiety and melancholia, neither produced extensive accounts of any mental illness. Several physicians practicing in Rome during the first two centuries CE, including Celsus (ca. 30 CE), Aretaeus of Cappadocia (ca. 150 CE), and Soranus of Ephesus (ca. 100 CE) provided more in-depth discussions of mental illnesses, including anxiety.

2. Dodds ER (1951). *The Greeks and the Irrational.* University of California Press, Berkeley. Finlay MI (2002). *The World of Odysseus.* New York Review of Books Classics, New York. Kagan D (2010). *Thucydides: The Reinvention of History.* Viking, New York.

3. Aristotle (1980). *The Nicomachean Ethics* (Ross D, trans.). Oxford World Classics, New York, p. 49.

4. Aristotle (1991). *The Art of Rhetoric* (Lawson-Tancred HC, trans.). Penguin, New York, p. 153.

5. Graver MR (2007). *Stoicism and Emotion.* University of Chicago Press, Chicago, p. 96.

6. Homer (1990). *The Iliad* (Fagles R, trans.). Viking, New York, p. 350.

7. Later, the Roman philosopher Epictetus similarly observed in regard to fear, "Nothing else changes the complexion, or causes trembling, or sets the teeth chattering." *Discourses,* Book 2, *On Anxiety.* www.perseus.tufts.edu/hopper/text?doc= Perseus%3Atext%3A1999. 70r.

8. Konstan D (2006). *The Emotions of the Ancient Greeks.* University of Toronto Press, Toronto, p. 135.

9. Homer (1990). *The Iliad* (Fagles R, trans.). Viking, New York, p. 350.

10. Aristotle, *Nicomachean Ethics,* p. 51.

11. Aristotle, *Nichomachean Ethics,* p. 50.

12. Certain categories of people, as well, were considered to be prone to different degrees of fear. Aristotle asserted that older people "are cowards and fear everything in advance; for their disposition is the opposite of that of the young." Fears that could be unnatural among young people could be normal among the aged. *Art of Rhetoric,* p. 175.

13. Finlay, 2002, p. 66.

14. Epictetus (1944). *Discourses* (Higginson TW, trans.). Walter J Black, Roslyn, NY, pp. 117–119.

15. Aristotle, *Art of Rhetoric,* p. 156.

16. Graver, 2007, p. 96.

17. Epictetus (1916). *The Discourses of Epictetus including the Enchiridion* (Matheson PE, trans.). Oxford University Press, New York, p. 306.

18. Linforth IM (1946a). *The Corybantic Rites in Plato.* University of California Publications in Classical Philology, 13, no. 5, pp. 121–163. Aristotle, *The Art of Rhetoric,* p 156. Religion provided not only central sources of anxiety but also the ritual

purifications that could remedy it, using procedures that could pacify the gods. Oracles such as Delphi provided reassurance: "Out of his divine knowledge, Apollo would tell you what to do when you felt anxious or frightened" (Dodds, 1951, p. 75).

19. Linforth IM (1946b). *Telestic Madness in Plato.* University of California Publications in Classical Philology, 13, no. 6, p. 165.

20. Hippocrates (1923–1931). *Works of Hippocrates* (Jones WHS, Withington ET, eds. and trans.). Harvard University Press, Cambridge, vol. 1, p. 263. See also: Wallace ER (1994). Psychiatry and its nosology: A historico-philosophical overview. In Sadler J, Schwartz M, Wiggins O (eds.), *Philosophical Perspectives on Psychiatric Diagnostic Classification.* Johns Hopkins University Press, Baltimore, p. 48. Simon B (1978). *Mind and Madness in Ancient Greece.* Cornell University Press, Ithaca, p. 235.

21. Simon, 1978, p. 229.

22. Roccatagliata G (1986). *A History of Ancient Psychiatry.* Greenwood Press, New York, p. 172.

23. Galen quoted in Radden J (ed.) (2000). Introduction: From melancholic states to clinical depression. *The Nature of Melancholy: From Aristotle to Kristeva.* Oxford University Press, New York, p. 10.

24. Galen quoted in Jackson SW (1986). *Melancholia and Depression: From Hippocratic Times to Modern Times.* Yale University Press, New Haven, p. 42.

25. Aristotle, *Nichomachean Ethics*, p. 49.

26. Aristotle, *Nichomachean Ethics*, p. 53.

27. Aristotle, *Nichomachean Ethics*, p. 51.

28. Celsus quoted in Roccatagliata, 1986, p. 187.

29. Aretaeus quoted in Porter R (2002). *Madness: A Brief History.* Oxford University Press, New York, p. 45.

30. Lucretius. *On the Nature of Things.* Classics.mit.edu/Carus/nature_things .3.iii.html. Retrieved December 6, 2010. Also cited in Svendsen L (2007). *A Philosophy of Fear.* Reaktion Books, London, p. 31.

31. Galen quoted in Jackson, 1986, p. 42.

32. Simon, 1978.

33. Plato, *Ion*, 533d–536d, quoted in Linforth, 1946a, p. 137.

34. Roccatagliata, 1986, p. 136.

35. Posidonius quoted in Roccatagliata, 1986, pp. 142–143.

36. Andreas quoted in Roccatagliata, 1986, p. 135.

37. Caelius Aurelianus quoted in Roccatagliata, 1986, p. 242.

38. Rosen G (1968). *Madness in Society.* Harper Torchbooks, New York, p. 96.

39. Caelius Aurelianus quoted in Porter, 2002, p. 43.

40. Sigerist HE (1943). *Civilization and Disease.* University of Chicago Press, Chicago, p. 150. Porter R (1999). *The Greatest Benefit to Mankind: A Medical History of Humanity.* WW Norton, New York, pp. 55–62.

41. The concept of humors resonates with the theories of neurochemical imbalances that developed toward the end of the twentieth century.

42. Galen quoted in Roccatagliata, 1986, p. 200.

43. Pott H (2009). Emotions, phantasia and feeling in Aristotle's rhetoric. In Close E (ed.), *Greek Research in Australia*. Flinders University, Adelaide.

44. Cicero (1927). *Tuscularum Disputationum* (King JE, trans.). William Heinemann, London, p. 355.

45. Lewis A (1970). The ambiguous word "anxiety." *International Journal of Psychiatry* 9, pp. 62–79.

46. Aristotle (2000). Brilliance and melancholy. In Radden J (ed.) *The Nature of Melancholy: From Aristotle to Kristeva*. Oxford University Press, New York, p. 58 (emphasis in original).

47. Galen quoted in Jackson, 1986, p. 42.

48. Fortenbaugh WW (2006). Aristotle and Theophrastus on the emotions. In *Aristotle's Practical Side: On His Psychology, Ethics, Politics and Rhetoric*. Brill, Leiden, p. 100. Fortenbaugh WW (2002). *Aristotle on Emotion*, 2nd ed. Duckworth, London, pp. 69, 71.

49. Aristotle, *Nicomachean Ethics*, p. 24.

50. Aristotle, *Nicomachean Ethics*, p. 54, 1117a.

51. Finlay, 2002.

52. Epicurus (1964). *Letters and Principal Doctrines and Vatican Sayings* (Geer RM, trans.). Bobbs-Merrill, New York, p. 64.

53. McReynolds P (1985). Changing conceptions of anxiety: A historical review and a proposed integration. *Issues in Mental Health Nursing* 7, pp. 131–158.

54. Lucretius quoted in Greenblatt S (2011). *The Swerve: How the World Became Modern*. WW Norton, New York, p. 196.

55. This is William James's summary of section 20 of Epictetus's Enchiridion. See Richardson RD (2006). *William James: In the Maelstrom of American Modernism*. Houghton Mifflin, New York, p. 53.

56. Epictetus quoted in Graver, 2007, p. 87.

57. Rosen, 1968, p. 132.

58. Linforth, 1946a, p. 134.

59. Plato (1926). *Laws*, vii, 790d–791a (Bury RG, trans.). Loeb Classical Library, Cambridge. See Linforth, 1946a, pp. 129–130. Dodds, 1951, pp. 78, 80.

60. Simon, 1978, pp. 143–144.

Chapter 3. From Medicine to Religion—and Back

1. Bloch M (1961). *Feudal Society*. University of Chicago Press, Chicago. Huizinga J (1924). *The Waning of the Middle Ages*. St. Martin's Press, New York.

2. St. Augustine (2009). *Confessions*. Oxford World Classics, New York, 8.29. Much later, American psychologist William James echoed this belief: "Of course the sovereign cure for worry is religious faith." *Talks to Teachers on Psychology: And to Students on Some of Life's Ideals* (1899/1983). Harvard University Press, Cambridge, p. 118.

3. Porter R (1999). *The Greatest Benefit to Mankind: A Medical History of Humanity*. WW Norton, New York, pp. 106–112.

4. Avicenna (ca. 1170/2000). Black bile and melancholia. In Radden J (ed.), *The Nature of Melancholy: From Aristotle to Kristeva.* Oxford University Press, New York, p. 77.

5. Ficino M (1482/2000). *Three Books of Life.* In Radden J (ed.), *The Nature of Melancholy,* p. 88.

6. Platter quoted in Porter R (2002). *Madness: A Brief History.* Oxford University Press, New York, p. 52.

7. Shorter E (1992). *From Paralysis to Fatigue: A History of Psychosomatic Illness in the Modern Era.* Free Press, New York, p. 15. Glas G (2003). A conceptual history of anxiety and depression. In den Boer JA, Sitsen JMA (eds.), *Handbook on Depression and Anxiety: A Biological Approach.* Marcel Dekker, New York, pp. 1–44.

8. E.g., Jackson SW (1986). *Melancholia and Depression: From Hippocratic Times to Modern Times.* Yale University Press, New Haven. Radden J (2000). Introduction: From melancholic states to clinical depression. Porter, 2002.

9. Burton R (1621/2001). *The Anatomy of Melancholy.* (Jackson H, ed.). New York Review Books, New York, vol. I, p. 131.

10. Burton, 1621/2001, vol. I, p. 170.

11. Burton, 1621/2001, vol. I, p. 170.

12. Burton, 1621/2001, vol. I, pp. 261, 385, 388.

13. Burton, 1621/2001, vol. I, pp. 363, 386, 335.

14. Burton, 1621/2001, vol. III, pp. 142, 148.

15. Burton, 1621/2001, vol. I, pp. 261, 386, 387.

16. Burton, 1621/2001, vol. I, p. 337.

17. Burton, 1621/2001, vol. II, p. 243.

18. MacDonald M (1983). *Mystical Bedlam: Madness, Anxiety, and Healing in Seventeenth-Century England.* Cambridge University Press, London.

19. MacDonald, 1983, pp. 67, 76, 107.

20. MacDonald, 1983, pp. 88–89, 99, 41.

21. MacDonald, 1983, p. 187.

22. MacDonald, 1983, p. 158.

23. De Montaigne M (1958). *The Complete Essays of Montaigne.* Stanford University Press, Palo Alto, pp. 53, 52, 449.

24. De Montaigne, 1958, p. 449.

25. Hobbes cited in Fisher P (2002). *The Vehement Passions.* Princeton University Press, Princeton, p. 115.

26. Locke J (1996). *Some Thoughts Concerning Education and of the Conduct of Understanding* (Grant RW, Tarcov N, eds.). Hackett Publishing, Indianapolis, p. 88.

27. Saul H (2001). *Phobias: Fighting the Fear.* Arcade, New York, p. 27.

28. Hume D 1739–40/1978. *Treatise of Human Nature* (Selby-Bigge LA, ed.). Oxford University Press, Oxford. Book II, Part III, IX, pp. 438–448.

29. Hume, 1739–40/1978, pp. 439–440, 446.

30. Micale M (2008). *Hysterical Men: The Hidden History of Male Nervous Illness.*

Harvard University Press, Cambridge, p. 23. Shorter E (1997). *A History of Psychiatry: From the Era of the Asylum to the Age of Prozac.* Wiley, New York, pp. 22–23.

31. Willis T (1683/1971). *Two Discourses Concerning the Soul of Brutes.* Scholars' Facsimiles and Reprints, Gainesville, FL, p. 191.

32. Robinson quoted in Porter, 2002, p. 126.

33. Cheyne G (1733/1991). *The English Malady* (Porter R, ed.). Tavistock/Routledge, New York.

34. Scull A (2009). *Hysteria: The Biography.* Oxford University Press, New York, p. 57.

35. Cheyne, 1733/1991, p. 343. Also quoted in Scull, 2009, p. 50. See also Cheyne, pp. 260–261.

36. Porter, 1991, preface, p. xxxiii. In Cheyne, 1733/1991. See also pp. 194, 333.

37. Porter 1991, preface, p. ii.

38. Cheyne, p. 52. Porter, 1991, preface, p. xli.

39. Porter, 1991, p. xxxii.

40. Jackson, 1986, pp. 124–125. Porter, 2002, p. 128.

41. Jackson, pp. 126, 299.

42. Lewis A (1967). Melancholia: A historical review. In Lewis A, *The State of Psychiatry: Essays and Addresses.* Routledge & Kegan Paul, London, p. 78.

43. Jackson, 1986, p. 306.

44. Whytt quoted in Jackson, 1986, p. 297.

45. Battie W (1758/1962). *A Treatise on Madness.* Dawson's, London, p. 28.

46. Battie, 1758/1962, pp. 35–36, 90, 34.

47. Shorter, 1997, p. 22.

48. Porter, 1999, pp. 194, 269.

49. Shorter, 1992, p. 24.

50. Trotter T (1807). *A View of the Nervous Temperament.* Longman, London, p. xvii. Also quoted in Bynum WF (1985). The nervous patient in eighteenth- and nineteenth-century Britain: The psychiatric origins of British neurology. In Bynum WF, Porter R, and Sheperd M (eds.). *The Anatomy of Madness: Essays in the History of Psychiatry.* Tavistock, New York, p. 93.

51. Scull, 2009, p. 26.

52. Porter, 2002, pp. 128–130.

Chapter 4. The Nineteenth Century's New Uncertainties

1. Porter R (1999). *The Greatest Benefit to Mankind: A Medical History of Humanity.* WW Norton, New York, p. 314.

2. Goldstein J (2002). *Console and Classify: The French Psychiatric Profession in the Nineteenth Century.* Cambridge University Press, New York, p. 211.

3. Rosenberg CE (2007). *Our Present Complaint: American Medicine, Then and Now.* Johns Hopkins University Press, Baltimore, p. 118. Shorter E (1992). *From Paralysis to Fatigue: A History of Psychosomatic Illness in the Modern Era.* Free Press, New York, pp. 214–215.

4. Berrios GE, Link C (1995). Anxiety disorders: Clinical section. In Berrios GE, Porter R (eds.), *A History of Clinical Psychiatry: The Origin and History of Psychiatric Disorders.* New York University Press, New York, p. 549.

5. Oppenheim J (1991). *Shattered Nerves: Doctors, Patients, and Depression in Victorian England.* Oxford University Press, New York, p. 37.

6. Griesinger quoted in Shorter E (1997). *A History of Psychiatry: From the Era of the Asylum to the Age of Prozac.* Wiley, New York, p. 76. See also Berrios and Link, 1995. During the last third of the nineteenth century, degeneration theory, which posited that mental illness was not only inherited but also became worse as it was passed down through the generations, gained popularity. For the most part, however, nervous conditions such as anxiety were not thought to be inheritable and so were not considered to be degenerative.

7. Brown quoted in Scull A (2011). *Madness: A Very Short Introduction.* Oxford University Press, New York, p. 37.

8. Zeldin T (1981). *France, 1848–1945: Anxiety and Hypocrisy.* Oxford University Press, New York, pp. 20–21.

9. Darwin C (1872/2007). *The Expression of Emotions in Man and Animals,* 3rd ed. Filiquarian Publishing, Minneapolis, p. 354.

10. Darwin, 1872/2007, p. 43.

11. Darwin, 1872/2007, p. 365.

12. E.g., Nesse RM, Jackson ED (2006). Evolution: Psychiatric nosology's missing biological foundation. *Clinical Neuropsychiatry* 3, pp. 121–131. Horwitz AV, Wakefield JC (2007). *The Loss of Sadness: How Psychiatry Transformed Ordinary Sorrow into Depressive Disorder.* Oxford University Press, New York.

13. Shorter, 1992, p. 215.

14. Da Costa JM (1871). On irritable heart: A clinical study of a form of functional cardiac disorder and its consequences. *American Journal of Medicinal Science* 61, pp. 17–52.

15. Berrios and Link, 1995.

16. Hecker quoted in Shorter E (2005). *A Historical Dictionary of Psychiatry.* Oxford University Press, New York, p. 27.

17. Glas G (2003). A conceptual history of anxiety and depression. In den Boer JA, Sitsen JMA (eds.), *Handbook on Anxiety and Depression: A Biological Approach,* 2nd ed. Marcel Dekker, New York, p. 6.

18. Snaith RP (1968). A clinical investigation of phobias. *British Journal of Psychiatry* 114, p. 673. Berrios GE (1995b). The psychopathology of affectivity: Conceptual and historical aspects. *Psychological Medicine* 15, p. 269.

19. Maudsley quoted in Shorter, 1997, p. 90.

20. Berrios and Link, 1995, p. 549.

21. Kierkegaard S (1844/1980). *The Concept of Anxiety.* Princeton University Press, Princeton, p. xiii.

22. Kierkegaard S (1843/1992). *Either/Or.* Penguin, New York.

23. Gay P (2002). *Schnitzler's Century: The Making of Middle-Class Culture, 1815–1914*. WW Norton, New York.

24. Although Beard first used this term in an 1869 article, his renowned book did not appear until 1881. Beard C (1881). *American Nervousness: Its Causes and Consequences*, Putnam, New York.

25. Beard, 1881, p. 17.

26. Scull A (2009). *Hysteria: The Biography*. Oxford University Press, New York, p. 97. Micale M (2008). *Hysterical Men: The Hidden History of Male Nervous Illness*. Harvard University Press, Cambridge, pp. 156–157. Oppenheim, 1991, p. 96.

27. Beard, 1881, vii–viii.

28. Beard, 1881, p. 105.

29. Dubois quoted in Shorter, 1992, p. 221.

30. Charcot quoted in Goldstein, 2002, p. 335.

31. Rosenberg, 2007, p. 83. See also Lutz T (1991). *American Nervousness: 1903*. Cornell University Press, Ithaca.

32. Lutz, 1991.

33. Shorter, 1997, p. 113.

34. Beard quoted in Glas, 2003, p. 7.

35. Rosenberg C (1997). *No Other Gods: On Science and American Social Thought* (rev. ed.). Johns Hopkins University Press, Baltimore, p. 108.

36. Allbutt quoted in Oppenheim, 1991, p. 9.

37. Scull, 2009, p. 12.

38. Rosenberg, 2007, p. 5.

39. Rosenberg, 2007, p. 3.

40. Scull, 2009, p. 12.

41. Guislain quoted in Shorter, 2013, *How Everyone Became Depressed: The Rise and Fall of the Nervous Breakdown*. Oxford University Press, p. 57.

42. Rosenberg, 2007, pp. 64–65.

43. Glas, 2003, p. 5. Shorter, 2013, p. 67. Krishaber M (1873). *De la neveropathie cerebro-cardiaque*. Masson, Paris.

44. Hartenberg P (1901). *Les timides et la timidité*. F. Alcan, Paris.

45. Berrios, Link, 1995. Berrios GE (1995a). Obsessive-compulsive disorder: Clinical section. In Berrios GE, Porter R (eds.), *A History of Clinical Psychiatry: The Origin and History of Psychiatric Disorders*. New York University Press, New York, pp. 573–592.

46. Kraepelin E (1902). *Clinical Psychiatry*. Norwood Press, Norwood, OR, p. 50.

47. Goldstein, 2002, p. 332.

48. Janet quoted in Glas G (1996). Concepts of anxiety: A historical reflection on anxiety and related disorders. In Westenberg HGM, den Boer JA, Murphy DL (eds.), *Advances in the Neurobiology of Anxiety Disorders*. John Wiley & Sons, New York, pp. 1–19, quote on 11.

49. Janet quoted in Lewis A (1970). The ambiguous word "anxiety." *International Journal of Psychiatry* 9, p. 68.

50. Janet's works experienced a brief reemergence with the rise of the recovered memory movement in the 1980s, which valued his idea that memories of early childhood traumas could be repressed and then recovered through therapy many years after their occurrence.

51. Shorter, 1992, ch. 10.

52. Zeldin, 1981, p. 69.

53. Tuke DH (1894). Imperative ideas. *Brain* 17, pp. 179–197, quote on p. 179.

Chapter 5. The Freudian Revolution

1. As the previous chapter noted, Freud's contemporary Pierre Janet simultaneously developed a more comprehensive classification of anxiety conditions. However, Janet had few followers and so his contributions to the study of anxiety have not lasted.

2. Ellenberger H (1970). *The Discovery of the Unconscious*. Basic Books, New York, p. 488.

3. Nemiah JC (1985). Anxiety states (anxiety neuroses). In Kaplan HI, Sadlock BJ (eds.), *Comprehensive Textbook of Psychiatry*. Williams & Wilkins, Baltimore, p. 896.

4. Freud S (1894a/1959). The justification for detaching from neurasthenia a particular syndrome: The anxiety-neurosis. *Collected Papers*, vol. 1 (Riviere J, trans.). Basic Books, New York, pp. 76–106, quote on p. 77.

5. Freud S (1894b/1959). The defence neuro-psychoses. *Collected Papers*, vol. 1 (Riviere J, trans.). Basic Books, New York, p. 75.

6. Freud, 1894a/1959, p. 97. See also Makari G (2008). *Revolution in Mind: The Creation of Psychoanalysis*. Harper, New York, p. 88.

7. Freud S (1893/1959). On the psychical mechanism of hysterical phenomena. *Collected Papers*, vol. 1 (Riviere J, trans.). Basic Books, New York, p. 29. Freud, 1894a, p. 96.

8. Freud, 1894a/1959, p. 105.

9. American Psychiatric Association (1980). *Diagnostic and Statistical Manual of Mental Disorders*, 3rd ed. (*DSM-III*). American Psychiatric Association, Washington DC, p. 233. Later manuals turned GAD into a more specific worry-related disorder (see Chapter 8).

10. Freud, 1894a/1959, p. 79.

11. Freud, 1894a/1959, p. 80.

12. Freud, 1894a/1959, pp. 83–84.

13. Freud S (1896/1959). Heredity and the aetiology of the neuroses. *Collected Papers*, vol. 1 (Riviere J, trans.). Basic Books, New York, p. 143.

14. Ellenberger, 1970, pp. 534–546.

15. Freud, 1896/1959, p. 145 (italics in original).

16. Freud S (1905/1959). My views on the part played by sexuality in the aetiology of the neuroses. *Collected Papers*, vol. 1 (Riviere J, trans.). Basic Books, New York, p. 276.

17. Freud S (1898/1959). Sexuality in the aetiology of the neuroses. *Collected Papers*, vol. 1 (Riviere J, trans.). Basic Books, New York, p. 242.

18. Freud, 1905/1959, p. 282.

19. Freud, 1898/1959, p. 237.

20. Freud, 1894a/1959, p. 103.

21. Gay P (2002). *Schnitzler's Century: The Making of Middle-Class Culture, 1815–1914*. WW Norton, New York, pp. 46, 147–151.

22. While some scholars have speculated that Freud had an affair with his sister-in-law Minna Bernays, who lived with his family, most believe this was unlikely. However, the presence of another attractive female in the household could very well have exacerbated his sense of sexual repression.

23. See Makari, 2008, pp. 157–160.

24. Nemiah, 1985, p. 885.

25. Freud S (1936/1963). *The Problem of Anxiety*. WW Norton, New York, p. 20. Freud S (1926/1989). *Inhibitions, Symptoms and Anxiety*. WW Norton, New York, p. 63.

26. Freud, 1936/1963, pp. 113–114.

27. Freud, 1926/1989, p. 58.

28. Freud, 1936/1963, pp. 83–84, 88–89.

29. Freud, 1936/1963, p. 89.

30. Freud, 1936/1963, p. 90.

31. Freud, 1936/1963, p. 113.

32. Freud, 1926/1989, p. 32.

33. Freud, 1926/1989, p. 57.

34. Glas G (1996). Concepts of anxiety: A historical reflection on anxiety and related disorders. In Westenberg HGM, den Boer JA, Murphy DL (eds.), *Advances in the Neurobiology of Anxiety Disorders*. John Wiley & Sons, New York, pp. 3–19.

35. Ellenberger, 1970, p. 516.

36. Freud, 1926/1989, p. 75.

37. Shorter E (1997). *A History of Psychiatry: From the Era of the Asylum to the Age of Prozac*. Wiley, New York, p. 145. Hale NG (1995). *The Rise and Crisis of Psychoanalysis in the United States: Freud and the Americans, 1917–1985*. Oxford University Press, New York, p. 292. Freud himself did not believe that physicians should monopolize the practice of psychoanalysis. Especially in the United States, however, physicians obstinately maintained their exclusive right to practice analysis. They were successful until the 1970s, when the profession entered a precipitous decline.

38. Hale, 1995, p. 63. Freud visited the U.S. only once and did not have a warm view of the country, its culture, or its people.

39. Grinker R, Spiegel J (1945). *War Neuroses*. Blakiston, Philadelphia.

40. Horney K (1937). *The Neurotic Personality of Our Time*. WW Norton, New York, p. 22.

41. Horney, 1937, p. 54.

42. Horney, 1937, p. 79.

43. Fromm E (1941/1969). *Escape from Freedom*. Avon Books, New York, p. 318.

44. Shorter, 1997, p. 187.

Chapter 6. Psychology's Ascendance

1. Although behaviorists emphasized the role of environmental stimuli in causation, they took an individualist approach to theory and therapy. Their conception of the environment was devoid of social and cultural factors.

2. Ward R (2002). *Modernizing the Mind: Psychological Knowledge and the Remaking of Society*. Praeger, Westport, CT, ch. 2.

3. Capshew JH (1999). *Psychologists on the March: Science, Practice and Professional Identity in America, 1929–1969*. Cambridge University Press, Cambridge, pp. 1, 15–16, 29.

4. Herman E (1996). *The Romance of American Psychology*. University of California Press, Berkeley, pp. 100–112.

5. Boring EG, Van de Water M (1943). *Psychology for the Fighting Man*. National Research Council, Washington, DC, pp. 298–299.

6. Janis JL (1951). *Air War and Emotional Stress*. McGraw-Hill, New York. Rachman SJ (1990). *Fear and Courage*, 2nd ed. WH Freeman, New York, pp. 19–25, 38. Bourke J (2005). *Fear: A Cultural History*. Counterpoint, New York, p. 248.

7. Capshew, 1999, p. 243. Herman, 1996, p. 2.

8. Capshew, 1999, p. 4.

9. Danziger K (1990). *Constructing the Subject: Historical Origins of Psychological Research*. Cambridge University Press, New York. See also Richardson RD (2006). *William James: In the Maelstrom of American Modernism*. Houghton Mifflin, New York.

10. James W (1884). What is an emotion? *Mind* 9, pp. 188–205.

11. Hall quoted in Bourke, 2005, p. 116.

12. Watson JB (1913). Psychology as the behaviorist views it. *Psychological Review* 20, pp. 158–177, quote on p. 158.

13. Watson JB (1924). *Behaviorism*. People's Institute, Chicago, p. 104.

14. Watson JB, Rayner R (1920). Conditioned emotional reactions. *Journal of Experimental Psychology* 3, pp. 1–14. Contrary to Watson's assertion that Albert was a normal child, in fact he suffered from severe neurological impairments that likely stemmed from congenital hydrocephalus and mental retardation. See Fridlund AJ, Beck HP, Goldie WD, Irons G (2012). Little Albert: A neurologically impaired child. *History of Psychology*, advance online publication. DOI: 10.1037/a0026720.

15. Watson, Rayner, 1920, p. 14.

16. Beck HP, Levinson S, Irons G (2009). Finding Little Albert: A journal to John B. Watson's infant laboratory. *American Psychologist* 64, pp. 605–614.

17. Salter A (1949). *Conditioned Reflex Therapy: The Direct Approach to the Reconstruction of Personality*. Creative Age Press, New York, p. 100.

18. Quoted in Morawski JG (1997). Educating the emotions: Academic psychology, textbooks, and the psychology industry, 1890–1940. In Pfister J, Schnog N (eds.) *Inventing the Psychological: Toward a Cultural History of Emotional Life in America*. Yale University Press, New Haven, p. 232.

19. Mowrer OH (1939). A stimulus-response analysis of anxiety and its role as a reinforcing agent. *Psychological Review* 46, pp. 553–564. Although most research during this period followed Watson's strict learning approach, a few behaviorists tried to incorporate analytical insights. Most notably, in 1950 psychologist Neil Miller and sociologist John Dollard published *Personality and Psychotherapy: An Analysis in Terms of Learning, Thinking, and Culture* (McGraw-Hill, New York). Dollard and Miller reformulated psychoanalytic and social scientific concepts into the terms of learning theory, a far richer approach than was typical at the time, albeit one that had little subsequent influence.

20. Skinner BF (1953). *Science and Human Behavior*. Macmillan, New York.

21. Skinner BF (1971). *Beyond Freedom and Dignity*. Knopf, New York.

22. Seligman M (1970). On the generality of the laws of learning. *Psychological Review* 77, pp. 406–418. Seligman M (1971). Phobias and preparedness. *Behavior Therapy* 2, pp. 307–320.

23. Kagan J, Reznick, S, Snidman N (1988). Biological basis of childhood shyness. *Science* 240, pp. 167–171. Kagan J, Snidman N (2004). *The Long Shadow of Temperament*. Belknap Press, Cambridge, MA.

24. Jones MC (1924a). The elimination of children's fears. *Journal of Experimental Psychology* 7, pp. 382–390. See also Jones MC (1924b). A laboratory study of fear: The case of Peter. *Pedagogical Seminary* 31, pp. 308–315.

25. Wolpe J (1958). *Psychotherapy as Reciprocal Inhibition*. Stanford University Press, Palo Alto, pp. 201, 83.

26. Note that relaxation is an internal process that is seemingly incongruent with the type of observable, external phenomena on which behavioral theory focused.

27. Rachman, 1990, pp. 103–117.

28. Eysenck HJ (1980). Psychological theories of anxiety. In Burrows GD, Davies BM (eds.), *Handbook of Studies on Anxiety*. Elsevier/North-Holland Biomedical Press, New York, p. 21.

29. Rose N (1999). *Governing the Soul: Of the Private Self*, 2nd ed. Free Association Books, London, p. 237.

30. Eysenck HJ (1952). The effects of psychotherapy: An evaluation. *Journal of Consulting Psychology* 16, pp. 319–324. Eysenck HJ (1965). The effects of psychotherapy. *International Journal of Psychiatry* 1, pp. 99–144.

31. Wolpe J (1988). *Life Without Fear: Anxiety and Its Cure*. New Harbinger Publications, Oakland, p. 114.

32. Weitzman B (1967). Behavior therapy and psychotherapy. *Psychological Review* 74, pp. 300–317.

33. Karzin AE (1975). *History of Behavior Modification: Experimental Foundations of Contemporary Research*. National Research Council, Washington, DC, pp. 197–204.

34. Fancher RT (1995). *Cultures of Healing: Correcting the Image of American Mental Health Care*. WH Freeman, New York.

35. Ellis A (1962). *Reason and Emotion in Psychotherapy*. Lyle Stuart, New York.

36. Beck AT (1976). Cognitive therapy of anxiety. Reprinted in Sahakian WS (ed.) *Psychopathology Today*, 3rd ed. FE Peacock, Itasca, IL, p. 97.

37. Among the studies are: Abramowitz J (1997). Effectiveness of psychological and pharmacological treatments for obsessive-compulsive disorder: A quantitative review. *Journal of Consulting and Clinical Psychology* 65, pp. 44–52. Fedoroff IC, Taylor S (2001). Psychological and pharmacological treatments of social phobia: A meta-analysis. *Journal of Clinical Psychopharmacology* 21, pp. 311–324. Mitte K (2005). Meta-analysis of cognitive-behavioral treatments for generalized anxiety disorder: A comparison with pharmacotherapy. *Psychological Bulletin* 131, pp. 785–795. Hunot V, Churchill R, Teixeira V, Silva de Lima M (2007). Psychological therapies for generalized anxiety disorders. *Cochrane Database of Systematic Reviews* 1, art. no. CD001848. DOI: 10.1002/14651858.CD001848.pub4.

38. See, among others: Mojtabai R, Olfson M (2010). National trends in psychotropic medication polypharmacy in office-based psychiatry. *Archives of General Psychiatry* 67, pp. 26–36. Mojtabai R, Olfson M (2008). National trends in psychotherapy by office-based psychiatrists. *Archives of General Psychiatry* 65, pp. 962–970. Goisman RM, Warshaw MG, Keller MB (1999). Psychosocial treatment prescriptions for generalized anxiety disorder, panic disorder, and social phobia, 1991–1996. *American Journal of Psychiatry* 156, pp. 1819–1821. Blanco C, Olfson M, Stein DJ, Simpson HB, Gameroff MJ, Narrow WII (2006). Treatment of obsessive-compulsive disorder by U.S. psychiatrists. *Journal of Clinical Psychiatry* 67, pp. 946–951.

Chapter 7. The Age of Anxiety

1. May R (1950/1977). *The Meaning of Anxiety*. Pocket Books, New York, p. 3.

2. See Barrett W (1958). *Irrational Man: A Study in Existential Philosophy*. Anchor Books, New York, p. 226.

3. Freud quoted in May, 1950/1977, epigraph.

4. Auden WH (1947/2011). *The Age of Anxiety*. Princeton University Press, Princeton.

5. Friedan B (1963/2001). *The Feminine Mystique*. WW Norton, New York, p. 407.

6. Hale NG (1995). *The Rise and Crisis of Psychoanalysis in the United States: Freud and the Americans, 1917–1985*. Oxford University Press, New York, p. 289.

7. See especially Herzberg D (2009). *Happy Pills in America: From Miltown to Prozac*. Johns Hopkins University Press, Baltimore. Tone A (2009). *The Age of Anxiety: A History of America's Turbulent Affair with Tranquilizers*. Basic Books, New York.

8. Kline quoted in Tone, 2009, p. 52. See also: Shorter E (2007). *A History of*

Psychiatry: From the Era of the Asylum to the Age of Prozac. Wiley, New York. Healy D (1997). *The Antidepressant Era*. Harvard University Press, Cambridge.

9. Berger FM (1970). Anxiety and the discovery of the tranquilizers. In Ayd FJ, Blackwell B (eds.), *Discoveries in Biological Psychiatry*. JB Lippincott, Philadelphia, pp. 115–129, quotes on pp. 120, 127.

10. Tone, 2009. Herzberg, 2009.

11. Tone, 2009, p. 75.

12. Greene J, Herzberg D (2010). Hidden in plain sight: The popular promotion of prescription drugs in the 20th century. *American Journal of Public Health* 100, pp. 793–803. Herzberg, 2009, p. 81.

13. Tone, 2009, p. 27. Herzberg 2009, pp. 44, p. 31, p. 37. Metzl JM (2003). *Prozac on the Couch: Prescribing Gender in the Era of Wonder Drugs*. Duke University Press, Durham.

14. Rickels K (1968). Drug use in outpatient treatment. *American Journal of Psychiatry* 125, pp. 20–31, quote on p. 10.

15. Blackwell B (1973). Psychotropic drugs in use today: The role of diazepam in medical practice. *JAMA* 225, pp. 1637–1641.

16. Tone, 2009, p. 90.

17. Tone, 2009, p. 153. Smith MC (1985). *Small Comfort: A History of the Minor Tranquilizers*. Praeger, New York, pp. 31–32. Herzberg, 2009, p. 138.

18. Herzberg, 2009, p. 82.

19. Herzberg, 2009, p. 128. Tone, 2009, pp. 156, 137. Smith, 1985, pp. 187, 120.

20. Quoted in Smith, 1985, p. 91.

21. Ayd FJ (1970). The impact of biological psychiatry. In Ayd FJ, Blackwell B (eds.), *Discoveries in Biological Psychiatry*. JB Lippincott, Philadelphia, p. 243.

22. Tone A (2005). Listening to the past: History, psychiatry, and anxiety. *Canadian Journal of Psychiatry* 50, p. 378.

23. Report quoted in Tone, 2009, p. 86.

24. Burney quoted in Smith, 1985, p. 73.

25. Klerman G (1970). Drugs and social values. *International Journal of the Addictions* 5, pp. 313–319.

26. Edwards quoted in Smith, 1985, p. 189.

27. Gabe J, Bury M (1991). Tranquillisers and health care in crisis. *Social Science and Medicine* 32, 449–454. Gabe J (1990). Towards a sociology of tranquillizer prescribing. *British Journal of Addiction* 85, 41–48. Tone, 2009.

28. Blackwell, 1973, p. 1640. See also Rickels K (1978). Use of antianxiety agents in anxious outpatients. *Psychopharmacology* 58, 1–27.

29. Ayd, 1970, p. 239.

30. Tone, 2009, p. 203. Smith, 1985, p. 32.

31. Grob GN (1990). World War II and American psychiatry. *Psychohistory Review* 19, pp. 41–69. Grob GN (1991). Origins of DSM-I: A study of appearance and reality. *American Journal of Psychiatry* 148, pp. 421–431.

32. American Psychiatric Association (1952). *Diagnostic and Statistical Manual of Mental Disorders*, American Psychiatric Association, Washington, DC, p. 31.

33. American Psychiatric Association (1968). *Diagnostic and Statistical Manual of Mental Disorders*, 2nd ed. American Psychiatric Association, Washington, DC, p. 39.

34. Langner TS (1962). A twenty-two item screening score of psychiatric symptoms indicating impairment. *Journal of Health and Social Behavior* 3, pp. 269–276.

35. Menninger quoted in Porter R (2002). *Madness: A Brief History*. Oxford University Press, New York, p. 208.

36. Rosenberg CE (2007). *Our Present Complaint: American Medicine, Then and Now*. Johns Hopkins University Press, Baltimore. Szasz TS (1961/1974). *The Myth of Mental Illness*. Hoeber-Harper, New York. Rosenhan DL (1972). On being sane in insane places. *Science* 179, pp. 250–258. Cooper J, Rendell R, Burland B, Sharpe L, Copeland J, Simon R (1972). *Psychiatric Diagnosis in New York and London*. Oxford University Press, London.

37. Feighner JP, Robins E, Guze SB, Woodruff RA, Winokur G, Munoz R (1972). Diagnostic criteria for use in psychiatric research. *Archives of General Psychiatry* 26, pp. 57–63. The other eleven conditions were primary and secondary affective disorders, schizophrenia, hysteria, antisocial personality disorder, alcoholism, drug dependence, mental retardation, organic brain syndrome, homosexuality, transsexualism, and anorexia nervosa.

38. Klein DF (1981). Anxiety reconceptualized. In Klein DF, Rabkin J (eds.), *Anxiety: New Research and Changing Concepts*. Raven Press, New York, pp. 235–263, quote on p. 242.

39. E.g., Margraf J, Ehlers A, Roth WT (1986). Biological models of panic disorder and agoraphobia: A review. *Behavior Research and Therapy* 24, p. 556. Rickels K, Rynn MA (2001). What is generalized anxiety disorder? *Journal of Clinical Psychiatry* 62, pp. 4–11.

40. Quoted in Bayer R, Spitzer RL (1985). Neurosis, psychodynamics, and DSM-III: A history of the controversy. *Archives of General Psychiatry* 42, p. 189.

41. Spitzer RL (1978). The data-oriented revolution in psychiatry. *Man and Medicine* 3, pp. 193–194.

42. American Psychiatric Association, 1968, p. 39.

43. Barlow DH (1988). *Anxiety and Its Disorders*. Guilford Press, New York, p. 567.

44. Eysenck HJ (1947). *Dimensions of Personality*. Routledge & Kegan Paul, London. Eysenck HJ, Wakefield JA, Friedman AF (1983). Diagnosis and clinical assessment: The DSM-III. *Annual Review of Psychology* 34, pp. 167–193. Healy, 1997, p. 257.

45. Klerman GL (1983). The significance of DSM-III in American psychiatry. In Spitzer RL, Williams JB, Skodol AE (eds.), *International Perspectives on DSM-III*. American Psychiatric Press, Washington, DC, pp. 3–24.

46. Baldessarini RJ (2000). American biological psychiatry and psychopharma-

cology, 1944–1994. In Menninger RW, Nemiah JC (eds.), *American Psychiatry after World War II*. American Psychiatric Press, Washington, DC, pp. 371–412.

47. American Psychiatric Association (1980). *Diagnostic and Statistical Manual of Mental Disorders*, 3rd ed. American Psychiatric Association, Washington, DC, p. 225.

48. Curtis GC, Magee WJ, Eaton WW, Wittchen HU, Kessler RC (1998). Specific fears and phobias: Epidemiology and classification. *British Journal of Psychiatry* 173, p. 213. Kessler RC, Stein MB, Berglund P (1998). Social phobia subtypes in the National Comorbidity Survey. *American Journal of Psychiatry* 155, p. 614.

49. Healy, 1997, p. 237. Mojtabai R, Olfson M (2010). National trends in psychotropic medication polypharmacy in office-based psychiatry. *Archives of General Psychiatry* 67, pp. 26–36. Mojtabai R, Olfson M (2008). National trends in psychotherapy by office-based psychiatrists. *Archives of General Psychiatry* 65, pp. 962–970.

50. Kessler RC, Koretz D, Merikangas KR, Wang PS (2004). The epidemiology of adult mental disorders. In Levin BL, Petrilice J, Hennessey KD (eds.), *Mental Health Services: A Public Health Perspective*. Oxford University Press, New York, pp. 157–176, quote on p. 165.

51. Staub ME (2011). *Madness as Civilization: When the Diagnosis Was Social, 1948–1980*. University of Chicago Press, Chicago. Kessler RC, Foster CL, Saunders WB, Stang PE (1995). The social consequences of psychiatric disorders: Part I. Education attainment. *American Journal of Psychiatry* 54, pp. 313–321. Kessler RC, Berglund PA, Foster CL, Saunders WB, Stang PE, Walters EE (1997). The social consequences of psychiatric disorders: Part II. Teenage childbearing. *American Journal of Psychiatry* 154, pp. 1405–1411.

52. Tone, 2009, p. 212. Healy, 1997, p. 196. http://psychcentral.com/lib/2010/top–25-psychiatric-prescriptions-for–2009. Accessed January 16, 2012.

53. Kramer PD (1993). *Listening to Prozac: A Psychiatrist Explores Antidepressant Drugs and the Remaking of the Self*. Viking, New York.

54. Kramer, 1993, p. 247.

55. *Newsweek* 1994, February 7.

56. Kramer, 1993, p. 175.

57. Healy D (1991). The marketing of 5-Hydroxytryptamine: Depression or anxiety? *British Journal of Psychiatry* 158, p. 739.

58. Horwitz AV (2010). How an age of anxiety became an age of depression. *Milbank Quarterly* 88, pp. 112–138.

59. Quoted in Tone, 2009, p. 217.

60. Tone, 2009, p. 220.

61. See the many reviews in the "anxiety" category in the Cochrane database: http://www2.cochrane.org/reviews/. See also Mullins CD, Shaya FT, Meng F, Wang J, Harrison D (2005). Persistence, switching, and discontinuation rates among patients receiving sertraline, paroxetine, and citalopram. *Pharmacotherapy* 25, pp. 660–667.

62. Substance Abuse and Mental Health Services Administration (2012). *Mental Health, United States, 2010.* Department of Health and Human Services, Rockville, MD, p. 138.

63. Porter R (1997). *The Greatest Benefit to Mankind: A Medical History of Humanity.* WW Norton, New York, p. 521.

Chapter 8. The Future of Anxiety

1. Svendson quoted in Fisher P (2002). *The Vehement Passions.* Princeton University Press, Princeton, p. 116.

2. Pinker S (2011). *The Better Angels of Our Nature: Why Violence Has Declined.* Viking, New York.

3. Kessler RC, Chiu WT, Demler O, Walters EE (2004). Prevalence, severity, and comorbidity of 12-month DSM-IV disorders in the National Comorbidity Survey Replication. *Archives of General Psychiatry* 62, 617–627. American Psychiatric Association (1980). *Diagnostic and Statistical Manual of Mental Disorders,* 3rd ed. American Psychiatric Association, Washington, DC, p. 225.

4. Shorter E (1992). *From Paralysis to Fatigue: A History of Psychosomatic Illness in the Modern Era.* Free Press, New York, p. 296.

5. Kramer PD (1993). *Listening to Prozac: A Psychiatrist Explores Antidepressant Drugs and the Remaking of the Self.* Viking, New York, p. 296.

6. LeDoux J (2002). *Synaptic Self: How Our Brains Become Who We Are.* Penguin, New York, p. ix.

7. Alfonso CA (2012). Letter to the editor. *New York Times Magazine,* April 22, p. 8.

8. Vaughan SC (1997). *The Talking Cure: The Science Behind Psychotherapy.* Grosset/Putnam, New York, p. 147.

9. Barber C (2008). *Comfortably Numb: How Psychiatry Is Medicating a Nation.* Pantheon, New York, p. 167.

10. MacFarquhar L (2007). Two heads: A marriage devoted to the mind-body problem. *The New Yorker,* February 12, p. 69.

11. Buckley C (2009). In the tents, most faces feature model pout. *New York Times,* February 10, A17.

12. Gall FJ (1835). *On the Organ of the Moral Qualities and Intellectual Faculties, and the Plurality of the Cerebral Organs.* Marsh, Capen & Lyon, Boston, pp. 45–46.

13. Smoller J (2012). *The Other Side of Normal: How Biology Is Providing the Clues to Unlock the Secrets of Normal and Abnormal Behavior.* William Morrow, New York, p. 334.

14. In 2006 various psychodynamically oriented groups collaborated to produce an alternative diagnostic manual. This manual, however, has not had any evident impact on mental health practice. Alliance of Psychodynamic Organizations (2006). *Psychodynamic Diagnostic Manual,* 1st ed. Psychodynamic Diagnostic Model, Bethesda, MD.

15. Beginning with the *DSM-5*, the title will employ arabic rather than roman numerals, to more easily accommodate presumed frequent revisions to the manual.

16. Kupfer DJ, Kuhl EA, Narrow WH, Regier DA (2009). On the road to DSM-V. http://www.dana.org/news/cerebrum/detail.aspx?id=23560. A more practical rationale was the desire to make the categories in the *DSM* commensurate with the current International Classification of Diseases.

17. Greenberg G (2010). Inside the battle to define mental illness. *Wired*, December 27, http://www.wired.com/magazine/2010/12/ff_dsmv/all/1.

18. Greenberg, 2010.

19. Schatzberg AF, Scully JH, Kupfer DJ, Regier DA (2009). Setting the record straight: A response to Frances commentary. *Psychiatric Times* 26, July 1.

20. I do not here consider the changes recommended for the PTSD diagnosis. Although the manuals since the *DSM-III* had placed PTSD within the anxiety disorders category, the *DSM-5* work group suggested moving it to a self-contained category.

21. E.g., Kendler KS, Gardner CO (1998). Boundaries of major depression: An evaluation of DSM-IV criteria. *American Journal of Psychiatry* 155, pp. 172–177. Kessler RC, Merikangas KR, Berglund P, Eaton WW, Koretz DS, Walters EE. (2003). Mild disorders should not be eliminated from the *DSM-V*. *Archives of General Psychiatry* 60, pp. 1117–1122.

22. Regier DA, Narrow WE, Kuhl EA, Kupfer DJ (2009). The conceptual development of DSM-V. *American Journal of Psychiatry* 166, pp. 645–650, quote on p. 649.

23. Kupfer et al., 2009, p. 5.

24. Weinfurt KP, Cella D, Forrest CB, Shulman L (2011). NIH PROMIS 2011: Advancing PRO Science in Clinical Research and Patient Care. http://www.nihpromis.org/science/presentations. January 11.

25. Srole L, Langner TS, Michael ST, Kirkpatrick P, Opler MK, Rennie TAC (1962/1978). *Mental Health in the Metropolis: The Midtown Manhattan Study*, rev. ed., enlarged. McGraw-Hill, New York.

26. Burton R 1621/2001. *The Anatomy of Melancholy*. New York Review Books, New York, vol. I: 261.

27. Hippocrates (2010). *Aphorisms*. Kessinger Publishing, Whitefish, MN, Section 6.23.

28. Merikangas KR (1995). Contribution of genetic epidemiologic research to psychiatry. *Psychopathology* 28, p. 48. Stavrakaki C, Vargo B (1986). The relationship of anxiety and depression: A review of the literature. *British Journal of Psychiatry* 149, pp. 7–16. Kessler, Chiu, et al., 2004. Barlow DH (1988). *Anxiety and Its Disorders*. Guilford Press, New York. Ormel J, Oldehinkel AJ, Goldberg DP, Hodiamont PP, Wilmink FW, Bridges K, et al. (1995). The structure of common psychiatric symptoms: How many dimensions of neurosis? *Psychological Medicine* 25, pp. 521–530. Tyrer P (1985). Neurosis divisible. *Lancet*, March 23, pp. 685–688.

29. Merikangas KR et al. (1996). Comorbidity and boundaries of affective disorders with anxiety disorders and substance misuse: Results of an international task force. *British Journal of Psychiatry* 168, suppl. 30, p. 64. Zimmermann M, McDermut W, Mattia J (2002). Frequency of anxiety disorders in psychiatric outpatients with major depressive disorder. *American Journal of Psychiatry* 157, 1337–1340. Weissman MM, Merikangas KR (1986). The epidemiology of anxiety and panic disorders: An update. *Journal of Clinical Psychiatry* 47, pp. 11–17.

30. Kendler KS (1996). Major depression and generalized anxiety disorder: Same genes, (partly) different environments—revisited. *British Journal of Psychiatry* 30, pp. 68–75. Kendler KS, Gardner CO, Gatz M, Pedersen NL (2007). The sources of comorbidity between major depression and generalized anxiety disorder in a Swedish national twin sample. *Psychological Medicine* 37, pp. 453–462.

31. Healy D (1997). *The Antidepressant Era.* Harvard University Press, Cambridge, p. 175. Kramer, 1993, p. 289.

32. Regier et al., 2009, p. 646.

33. Carey B (2012). Psychiatry manual drafters back down on diagnoses. *New York Times*, May 9, A11. Frances A (2012). Wonderful news: *DSM-5* finally begins its belated and necessary retreat. http://www.psychiatrictimes.com/blog/frances/content/article/10168/2068571.

34. American Psychiatric Association (1952). *Diagnostic and Statistical Manual of Mental Disorders.* American Psychiatric Association, Washington, DC, p. 32. American Psychiatric Association (1968). *Diagnostic and Statistical Manual of Mental Disorders,* 2nd ed. American Psychiatric Association, Washington, DC, p. 38.

35. American Psychiatric Association, 1980, p. 233.

36. American Psychiatric Association (1987). *Diagnostic and Statistical Manual of Mental Disorders,* 3rd ed., revised. American Psychiatric Association, Washington, DC, p. 252.

37. American Psychiatric Association (2013). *Diagnostic and Statistical Manual of Mental Disorders,* 5th ed. American Psychiatric Association, Washington, DC.

38. American Psychiatric Association, 1952, p. 32. American Psychiatric Association, 1968, p. 38.

39. Shorter E (2007). *A History of Psychiatry: From the Era of the Asylum to the Age of Prozac.* Wiley, New York, p. 292.

40. As noted in Chapter 1, the NIMH is developing an alternative diagnostic system, the RDoC, which would serve research as opposed to clinical purposes.

41. Oppenheim J (1991). *Shattered Nerves: Doctors, Patients, and Depression in Victorian England.* Oxford University Press, New York, p. 315.

INDEX

aerophobia, 27
Age of Anxiety, 119
agoraphobia, 2, 42, 62, 69, 111, 134
Albert B, 105–6
Allbutt, T. Clifford, 67
American Psychiatric Association, 127, 149–50
American Psychological Association, 102
Anatomy of Melancholy, The (Burton), 39–42
Andreas of Charystos, 27
animal spirits, 49
antidepressants, 139–40
anxiety, 2, 4–5, 16–18, 47, 98–99, 134–35, 143–44; behavioral approach to, 13, 108–9; behaviors related to, 5; brain activity associated with, 3, 8–9, 59, 145–46; Classical-era causes of, 20, 28–30; classification of, 48–49, 53, 68, 69–73; cognitive approaches to, 13, 108–9; culture and, 4–6, 92–93; depression and, 72 (*see also* comorbidity; depression); dimensional assessment of, 150–53, 158; disordered, 44–45; drugs for treatment of, 11–13, 95 (*see also* benzodiazepines; drug therapy; psychopharmacology; tranquilizers; *and individual drugs*); dysfunctional, 1, 4, 7; economics and, 45–46, 51, 94; environmental influences on, 8, 10–11, 110; foundational condition of neuroses, 89–90; Freud's early writings on, 77–81; Hippocratic treatments for, 42; interactional networks and, 92; linked to broader psychotic states, 71; love and, 40; medicalization of, 55, 57, 58, 80; medical treatment for, 53; Napier's treatments for, 44; as nervous disorder, 50; in philosophy, 45–48; physiological aspects of, 1, 3, 58–64; as

political topic, 96; as primary symptom, 28; prominence of, in Western culture, 118–19; psychic aspects of, 62; religion as cause of, 37; repression and, 88–90; screening for, 138–39; severity of, 151; sexual causes for, 82; social circumstances and, 10, 50–51, 119, 139; spirituality and, 17, 63–64; subjective states of, 63; suppression of, 32; symptoms of, 50; threat and, 14, 87; transforming Freud's concept of, 85–88; treatments for, 2, 53, 83 (*see also* drugs; psychotherapy; therapy, behavioral); types of, 6–7, 13, 14, 15, 53, 87–88; underlying force of, 78–79
anxiety attacks, 80
anxiety disorders, 24–28, 78, 81–85
anxiety neurosis, 70
anxiolytics, 139–40
anxious expectation, 80, 156
anxiousness, temperament and, 9–10
Aretaeus of Cappadocia, 25–26, 166n1
Aristotle, 9, 14, 20, 21, 25, 30, 31
Association for Advancement of Behavior Therapy, 113
Augustine, 37, 80
Auster, Paul, 6–7
Avicenna, 11, 38
avoidance reactions, 107
Ayd, Frank, 127

Bacon, Francis, 48
Barlow, David, 135
Battie, William, 53
Beard, George, 51, 64–67, 76, 77, 119
Beck, Aaron, 114–15
behavioral psychology, 10, 47, 97–99, 102–9, 113